Safety Insights

Safety Insights

Success and Failure Stories of Practitioners

Edited by
Nektarios Karanikas
Maria Mikela Chatzimichailidou

A PRODUCTIVITY PRESS BOOK

First published 2021
by Routledge
6000 Broken Sound Parkway NW, Suite 300, Boca Raton, FL 33487-2742

and by Routledge
2 Park Square, Milton Park, Abingdon, Oxon, OX14 4RN

Routledge is an imprint of the Taylor & Francis Group, an informa business

ISBN: 978-0-367-44572-0 (hbk)
ISBN: 978-1-003-01077-7 (ebk)

Typeset in Garamond
by Deanta Global Publishing Services, Chennai, India

We dedicate this book to all workers, supervisors, managers, consultants and staff of associations, agencies and regulators who contribute to public, occupational, product and services safety.

The book is dedicated to everyone who is open to learning from the positive and negative experiences of others without judgement and is keen to share successful and unsuccessful safety stories with the social and professional communities.

Contents

Preface

A Few Lines from Nektarios

Why this book? To be honest, my first intention was to write a monograph, and the project of an edited book like the one you hold in your hands now was not in my to-do list until early 2019. Not that I have abandoned the idea of writing a book; I already started typing my chaotic and unordered thoughts, ideas and perspectives, but it will take time to complete this process, if ever! Meanwhile, I wondered why any reader would be interested in lines written by me only and not a collection of different, divergent or convergent voices about safety journeys? During those endoscopic moments, some random and scattered thoughts when attending and presenting industry and scientific conferences and reading or conducting research surfaced to my mind.

Considering safety conferences, I had noticed various interesting facts. First, who was presenting? Respected safety professionals, mainly safety managers and officers, curious and creative scholars who had something new to share about safety or, occasionally, teams consisting of industry and/or academics. All good, but were there any other voices that could be heard and did not have similar opportunities due to lack of time, limited funds, busy agendas, inadequate support from employers and so on? Do we really get all possible benefits of disseminating safety knowledge widely when sharing is typically restricted to the ones who are in the advantageous position to present and communicate their safety initiatives and results? Indeed, you could argue that there are other avenues to share, such as professional magazines, online articles, social media and blogs and so on. However, my experience suggests that the people involved in activities like the ones I listed above are more or less the same persons who present at conferences! On a side note, most of the times I have submitted a proposal to present at an industry conference, I have been asked to declare my experience in presenting along with all my academic and professional credentials so that the organisers can be confident that I am skilled and qualified to share with others. Really? Are we getting close to becoming something like a safety-sharing guild? Is this a demonstration of the inclusion of diversity we claim we pursue?

Second, during safety conferences, two types of safety stories are mainly presented: successes of ours and the failures of others. Usually, nothing is about the shortcomings of ours and rarely something is about the success of others, the latter especially used to support the success of ours again and claim the birth and prospective value of a new best practice. Sincerely, sometimes, it feels like attending marketing events with a lack of deep reflection and transparency about the choices available, limitations and strengths of different options, preconditions and difficulties when applying any approach and so on. Most of the time, we have been great at promoting and celebrating our gains and are keen to reveal the unfortunate journeys of others. Even during academic conferences, others and I spend considerably more time to communicate the journey of our studies and their results, and we invest only one presentation slide in sharing the limitations and weaknesses of our work. We want to show confidence about the significance of our findings, and we deprive the audiences of critical information that could lead them to test, accept or reject our work. Are we afraid of our work being rejected in practice, maybe because we receive the reactions of others too personally?

Third, despite the above, where does the knowledge and information from safety events go? How is it used and who are the beneficiaries? The first-hand witnesses are the ones who attend the safety events. They can evaluate each presentation in real time or retrospectively (e.g., discussion with the speaker during networking time, downloading the presentation). They can raise questions, ask for clarifications, email the presenter and so on. No problem with that, but I am wondering what happens next. What are the criteria used by each professional to reject, partially accept and experiment, fully endorse and introduce to the organisation or blend any safety approach with others? Is it the confidence, voice pitch and persuasion skills of the presenter? Is it a good understanding of the content and the verification that it matches or not the organisational policies and visions? Is it the feeling of threat to established norms and practices within each company? Is it the reluctance or not to get out of our comfort zone and try something new and different? Are there resource constraints that discourage changes and render attendance at events a pleasant break from job commitments, an opportunity to network, a possible chance for personal development? Even if someone decides to try a new approach, is he/she asking for input from end-users, supervisors and line managers or is he/she enforcing changes top-down just because the new sounds good and there have been some good results elsewhere? Therefore, I am afraid that we do not really know what happens with the shared knowledge after safety conferences conclude.

Regarding our research and studies, we conduct interviews and focus groups to collect data. What do we do next? We apply thematic analyses and coding to find common, frequent and emerging messages and answer our research questions and test our hypotheses. All this is great and represents what we have agreed as a valid methodology. Nonetheless, are we missing the fine details when we focus on key messages while processing whole stories? Do these messages make the same sense when separated from the story and mixed with the words of others? Is it legitimate to filter the accounts of others for the sake of our studies? There are indeed publications that include exact quotes from interviewees. Still, these once more are used to confirm the picture the researchers derived and cannot offer a complete understanding of the whole story when they are isolated from the latter.

The reasons above were the ones that triggered in me the idea to invite everyone openly to share reflective stories about safety successes and failures. I am grateful to Mikela, my fellow editor of this book, for seeing the value of this project and supporting me in this first-of-its-kind venture. Together, we announced the idea openly to everyone regardless of position and role, industry sector, level of experience in writing and type of safety practices followed. We decided to include everyone interested in sharing with courage and reveal the pros and cons of his/her methods, things that worked or not. We promised everyone unconditional support in their writing journeys. We filtered nothing regardless of how traditional or innovative before the eyes of safety gurus and thinkers. We strongly believe that nothing is absolutely perfect; everything has some value; everyone has the right to share without being judged and labelled as "old" or "new" school of safety thinking.

Writing stories in this book was a journey of self-discovery for everyone, but most importantly, it was a journey which we took all together as the characters are real and so are their stories. Eighteen professionals from various fields (e.g., occupational safety, transport safety, system safety, and patient safety), diverse backgrounds (e.g., frontline employees, engineers, managers, consultants), different industry sectors (e.g., healthcare,

construction, aviation, rail, infrastructure, road transport, and process industry) and from around the world represent a mosaic of authors who built this
book chapter by chapter. When we had to decide the order of the chapters,
we came up with no valid criterion. Who was to judge if the story of someone should be before or after the account of others, first or last appearing
in the book? This would immediately cancel our positions and beliefs I
described above. Thus, we decided to order the chapters in ascending order
of the first name of each contributing author. We would like the authors to
be remembered with their first names, as exactly we call each other in our
daily activities, and the authors could be any one of us.

Mikela and I are grateful to all the authors who accepted the challenge. We learned so much from each one of you. You allowed us to reflect
even further on what we accept and reject. We want to thank you for your
patience with our comments on each version of your work. Sometimes, I was
afraid that you would withdraw because we pushed you intensively to reflect
and not just describe. Fortunately, your will to share with others honestly
overcame the efforts needed to write your stories. This is another great success worthy to share, isn't it? Our massive thanks also to the publisher, Taylor
& Francis Group, who saw the significance of this project and accepted the
risk to support it. Our appreciation to the endorsers of this book too, who
generously invested their time in reading various chapters and writing their
encouraging comments on the value and potential impact of this work.

Last but not least, our thanks to all readers who decided to navigate
through the stories unfolded in this book. We are confident that regardless of the motive behind and your current role, you will find these narratives highly interesting. There are plenty of useful messages and insights for
everyone; the curious member of the public who uses products and services
and wants to understand the safety challenges behind the scenes, the worker
who wants to comprehend reasons and obstacles behind various safety management initiatives, the manager who is ready to challenge his/her thinking
and approach in the light of experiences of others and the scholar who is
keen to discover new research avenues.

A Few Lines from Mikela

I have been fortunate to have played several roles in the preparation of
this book. As an editor, I am deeply grateful to the authors for having the
courage and generosity to feature fortunate and unfortunate periods and

moments of their careers. Special thanks to my fellow editor and long-standing friend, Nektarios, who came up with the idea, took the lead in editing this book and once again offered me the pleasure to work and learn next to him. As an author and speaking on behalf of all the contributing authors, in this very first section of the book, I consolidate and confess the feelings and thoughts as we experienced them during the past year that this book was evolving.

After the completion of each chapter, we asked the authors, Aikaterini, Anne-Louise, Athanasios, Badar, Connor, Derek, Dimitrios, Genovefa, Keith, Marion, Mark, Nikolaos, Sikder, Mohammad, Spyridon and Steve, to reflect on aspects of the writing experience. They were asked to share what they enjoyed the most, what challenges they faced, how past occurrences have affected their current practice, and why they would recommend to other professionals to share their stories. Almost everyone admitted that joy, dare, improvement, collaboration and boldness were the five top states experienced while putting together the pieces of the stories. This is what I am sharing with you below based on the consolidation of the perspectives of all authors. It is the collective voice of everyone who contributed to this book; I am just the narrator!

Being asked to translate a learning process into a narrative was both rewarding and challenging for everyone. Sometimes we dismiss work we have done previously, regardless of the positive or negative outcomes, although the latter rather stays imprinted in our minds for longer. Writing our chapters was an exercise that pushed us to see the value and the pitfalls in the work we undertook months, years or decades ago. We found that selecting particular situations and looking back to wins and losses reminded us of the wide variety of challenges we have encountered in our professional lives and beyond. The stories we share with the readers are parts of ourselves and authentic accounts of our experiences. Finding the time, ensuring a quiet environment and sitting down to start typing our stories was not the most straightforward task considering all the other commitments and the lack of experience in this style of writing. However, it enabled us to relive the enjoyment of rediscovering the value of skills obtained and relearn key lessons from successes and failures, of course, with the benefit of hindsight and under the lenses of our current knowledge and experiences.

The encouragement from the editors to avoid the use of highly technical language and prefer a casual writing style was another factor that pushed us outside our comfort zone. It is true that if we had used technical jargon per area covered in our chapters, our stories would look more like technical reports than authentic narratives of ours regardless of any good intention. Furthermore, approaching the stories from the perspective of the person we were and the role we held back then was a great challenge. We caught ourselves being quite judgemental towards our past decisions and actions, finding it difficult to keep a positive perspective and avoid excessive hindsight to draw useful conclusions by appreciating the journey each story unfolds and not only the outcomes. Sometimes it can be hard to hold the mirror up to yourself for a reality check, especially while revisiting an emotionally tough experience. We felt challenged when trying to present the roles and the details of the stories along with our reflections and making our accounts as simple and comprehensible to the reader as possible.

We consider this piece of work as a mutually beneficial opportunity for us who unburied and brought back to life forgotten wins and losses and for the readers who will hopefully discover that they are not alone. Everybody has experienced successes and failures, and there is much power when looking back, inside and deep into ourselves. During our professional lives, we realise that some stories repeat themselves in the sense that many similar situations exist even today. Certainly, nothing is exactly the same, and the only constant is change; however, the essential ingredients are still the same: human, technical and environmental elements and systems. However, looking back on the reasons for taking particular decisions gives us another point of reference to evaluate our decisions.

The comparison of what we think and do today with past decisions and outcomes is a useful tool in moderating the present and plotting the future. Have we changed since then? Have we appreciated all positives and negatives? Did we leverage our successes and learned from our pitfalls? Moreover, through the process of writing, we realised that we must take more ownership of our language and how we deliver thoughts and opinions verbally and non-verbally, in written or oral forms. Knowing that other people's views and experience might be different, how do we render ourselves transparent and crystal clear while, at the same time, understanding and valuing all well-intended perspectives? Most importantly, when reflecting, we were talking to ourselves, and this endoscopy revealed the difficulty of using proper language to express ourselves to ourselves!

Safety is a discipline that enables progress through constant attention to detail while maintaining a good understanding and overview of the whole picture. Our accounts show exactly this. Putting our experiences on a piece of paper helped us to read behind the details of the events and get the whole picture. For that reason, we enthusiastically invite others to communicate their safety stories to the external observer in a reflective, transparent and timely manner. Every professional should take the opportunity to review key steps in their careers, both positive and negative, as it helps to reinforce good behaviours and hone better tools for decision making in our current roles. We often do not take the time to reflect on our past experiences, how they have shaped us personally and our careers and how they might have influenced others.

When we invest time and effort in exploring why we acted in specific ways or headed down particular paths, it can consolidate things that worked well and put things that did not work so well into perspective. It can be healing when we share our failures with others. It also may help others to be able to gain an alternative perspective on their own situations when they compare these with the experience of other professionals and, therefore, collectively learn and participate in further innovations. All in all, everyone involved in safety and risk has stories to tell: high tales of disaster avoided, Darwin Award moments and miraculous saves by unwitting heroes. These are the stuff of legend that really bring safety management to life and they bring our theories firmly into the realm of the daily life experience.

Editors

Nektarios Karanikas

Nektarios is Associate Professor in the Health, Safety and Environment Discipline at the School of Public Health and Social Work, Faculty of Health at the Queensland University of Technology (QUT), Brisbane, Australia. He was awarded his doctorate in Safety and Quality Management from Middlesex University (UK) and holds an MSc in Human Factors and Safety Assessment in Aeronautics from Cranfield University (UK). Nektarios graduated from the Hellenic Air Force Academy (GR) as an aeronautical engineer and worked as an officer in the Hellenic Air Force (HAF) for more than 18 years before he resigned at the rank of Lt. Colonel in 2014.

While in the HAF, Nektarios served in various positions related to maintenance and quality management and accident prevention and investigations, and he was lecturer and instructor for safety and human factors courses, the majority of which he designed or substantially revised. In his previous appointment as Associate Professor of Safety and Human Factors at the Amsterdam University of Applied Sciences (NL), in addition to his teaching activities, he supervised students conducting their research projects during their placement at various companies, he delivered in-house and public masterclasses in safety, risk and investigation management and he led a four-year co-funded project about aviation safety metrics as well as smaller contract research and consultancy projects. Nektarios' willingness to bring closer academia and industry led to his initiative to launch the International Cross-industry Safety Conference in 2016, which has run annually since then.

He holds engineering, occupational health and safety, human factors and project management professional qualifications and has been an active member of various prestigious international and regional associations. Nektarios has published numerous academic journal articles, including papers in top-tier safety journals such as *Safety Science* and *Risk Analysis*, peer-reviewed conference papers and book chapters and has been invited to speak at many international and regional summits and workshops. He is a member of editorial boards and a regular reviewer of safety, human factors and aerospace-related journals, and he volunteers in various activities of professional bodies.

Maria Mikela Chatzimichailidou

Mikela is a Systems and Assurance Engineer at WSP (UK) and a Visiting Fellow at Cranfield University (UK) and Imperial College London (UK). She was awarded her PhD in Complex Systems Safety and Human Factors, as well as her MSc in Systems Engineering Management, from Democritus University of Thrace (GR) where she also did her MEng as a Mechanical Engineer. As a Postdoctoral Research Associate at the University of Cambridge (UK), Mikela worked on patient safety for a year. During her three years at Imperial College London (UK), Department of Civil Engineering, she led versatile projects ranging from system safety and human factors to complexity management and systems engineering in transportation and infrastructure. Mikela is a Chartered Engineer with the Institution of Engineering and Technology (IET) and a Fellow of the Safety and Reliability Society (SaRS). She also sits on the UK National Council of SaRS and leads the Outreach and Early Careers Committee.

Mikela identifies herself as a heretical researcher and consulting engineer bringing together experience and expertise from healthcare, aviation, rail and infrastructure and in the fields of system and safety assurance, resilience, human factors, system engineering, product engineering and project management. She has been leading research and consulting assignments for more than a decade both in academia for some of the world's top universities, and in industry for a world-class design and built environment consultancy. During her career, Mikela has worked in research and development, consultancy and design roles and as a project manager, leading others to deliver client objectives and cutting-edge academic research. Her offering is that she brings experience of both the academic and industrial worlds with the potential to bridge the theory–practice gap and draw the best from each.

Mikela has published a book on project and risk management, a best practice report with Innovate UK and the Knowledge Transfer Network (UK), 20 peer-reviewed papers with international journals and over 40 conference papers. She is a reviewer in numerous international journals, such as *Accident Analysis & Prevention*, *Safety Science* and *Risk Analysis*, and a member of scientific and technical committees of conferences.

Contributors

Dimitrios Chionis

Dimitrios is a PhD student of Psychology at the University of Bolton at New York College, Athens, Greece. He was awarded his MSc in Human Factors in Aviation from Coventry University (UK), and he studied BSc Psychology at the Aristotelian University of Thessaloniki, Greece. Dimitrios has been an accredited psychologist, coach and researcher on safety and human factors for more than seven years. Additionally, he has been accredited for various psychological assessment tools, served in positions related to personnel recruitment and accident investigation and he is an instructor for safety and human factors courses.

Steve Denniss

Steve's career has been devoted to ensuring new systems and operational practices provide measurable and demonstrable beneficial outcomes through precise and pragmatic attention to assurance processes. Mastering the fundamentals of assurance best practice from ten years in the defence industry, Steve's first railway project was the Docklands Light Railway upgrade leading the assurance process integrating rolling stock, signalling and operations. Providing systems expertise and leadership worldwide by delivering systems assurance on mainline, metro and light rail projects, his career spans nearly 40 years, the last 25 in the rail industry in the UK, Europe, North America and the Middle East.

Badar Farooq

Badar Farooq is a highly skilled Health, Safety, Environment and Sustainability Professional possessing consulting, engineering, construction and industrial experience in an integrated corporate responsibility function across industries ranging from energy, power and infrastructure to real estate. He has primarily resided and worked in the MENA region. Badar has also conducted, facilitated and published scientific research in Switzerland as part of his undergraduate environmental science degree from the American University of Sharjah. He has complemented his studies with a mini MBA from the PwC Academy Dubai.

Athanasios Galanis

Athanasios became an Air Traffic Controller (ATC) after he graduated from the Hellenic Air Force Academy in 2003 and attended all relevant vocational training. He has also completed Flight and Ground Safety and Human Performance in Military Aviation (HPMA) courses and he holds a Master's in Organisation and Administration. During his career, Athanasios has undertaken various roles, including Ground Safety Officer and ATC Trainer, Instructor, Assessor and Supervisor. He has participated in NATO, EUROCONTROL and National committees on air traffic and safety. Furthermore, Athanasios has joined as a leader and member in various working groups on operational risk management, emergency response planning and other safety initiatives.

Nikolaos Gkionis

Nikolaos is a highly experienced Safety, Security and Risk Executive with an extensive background in establishing and leading safety and risk governance in multinational companies and organisations within a diverse portfolio of industries, including malls and retail, hospitality, leisure and entertainment, real estate, marinas, casinos and major events. He has a proven record of highly demanding projects of developing safety, security, risk and emergency strategies, policies and programs and holds a wide range of relevant academic and vocational qualifications. Nikolaos has been a Fellow member of the Institute of Strategic Risk Management (ISRM) and has participated in conferences, panels, publications and articles across all aspects of the safety and security disciplines.

Sikder Mohammad Tawhidul Hasan

Sikder is a Mechanical Engineer with an MBA and Postgraduate Diploma in Information Technology. He completed a six-month course on Occupational Health and Safety (OHS) from the International Labour Organisation and recently earned his MSc in Risk and Safety Management from Aalborg University, Denmark. Sikder has been working as an Assistant Inspector

General (Safety) in the Department of Inspection for Factories and Establishments (DIFE) of Bangladesh since 2015. Also, he is a member of the committee developing a curriculum for the National Occupational Safety and Health Training and Research Institute, the first of its kind in Bangladesh. Sikder also contributed to the draft of the National Plan of Action on OHS.

Mohammad Tahidul Islam

Mohammad started his professional journey as a Safety Inspector in the Department of Inspection of Factories and Establishment (DIFE) in 2015. Right after the Rana Plaza incident in 2014, the garments sector of Bangladesh adopted drastic measures to ensure workplace safety, of which he has been a part and a first-hand witness. His scope of works includes inspecting safety aspects of factories, implementing government laws, motivating the factory owners and training personnel to comply with safety standard measures. Mohammad is an electrical engineer and has completed numerous professional training courses, including one in Occupational Health and Safety from Aalborg University in Denmark.

Keith Johnson

Keith holds nearly three decades of experience across infrastructure, mining, quarries, construction, processing, rental, pastoral, meat processing and Australian defence industries. Keith has many and varied tertiary qualifications in health, safety, environment, accident forensics and legal studies, and he has been voluntarily delivering lectures at the Queensland University of Technology (QUT) in accident investigations and forensics. He has won numerous awards, including the Leadership Excellence Award, Safety and Health Excellence Award and Safety Initiatives Award, among many other accolades. His hobbies include building and carpentry, he is a classic car enthusiast and he likes to contribute to local and national charities, such as Oxfam, where he completed a 100 km trail which raised over AU$7,000 for the charity.

Aikaterini Karakatsani

Aikaterini is an experienced aviation safety professional with a demonstrated history of working in the aviation industry. She is the Safety Manager for an aerodrome operator in Greece and has more than 15 years of experience working for aviation organisations. Aikaterini is skilled in operations management, safety and risk management, project management for airlines, ground service providers and airports. She is a certified Aviation Safety and Quality Auditor with a Master's in Business Administration focused on Project Management from Cardiff Metropolitan University.

Genovefa Kefalidou

Genovefa is an Assistant Professor in Human–Computer Interaction, School of Informatics at the University of Leicester. She has a strong multidisciplinary background including computer science, systems engineering, psychology and human factors. Her research interests include user/passenger experience (PAX), sensemaking big data, cognitive computing and user-centred design for technologies and services in intelligent mobility, safety, navigation and transport, utilising multimodal, mixed-reality technologies and artificial intelligence (AI). She employs mixed-methods to investigate human and system behaviour and performance within complex socio-technical contexts to design advanced decision-support tools. She has worked on national and international projects for enhancing transport infrastructure and passenger experience.

Marion Kiely

Marion is a Health and Safety Consultant at Upstream, which she founded in 2016, with a view to helping clients navigate complex challenges and become more resilient. She is also a Senior Consultant with Art of Work, Communications Coordinator for the Ireland South IOSH Committee, part-time Lecturer at University College Cork and a Cynefin Authorised Trainer with Cognitive Edge. Marion adopts a progressive, straight-talking approach, and is an advocate of anthro-complexity and safety differently approaches. She is particularly interested in the area of psychosocial risk factors and has co-developed a wellbeing pulse with Cognitive Edge intending to bring about positive change in this area.

Spyridon Markou

Spyridon (aka Spyros) has always been passionate about aviation safety. He is a specialised professional with experience in safety and quality management, regulatory compliance, flight and ground operations and aviation technology. Spyros has a proven track record of monitoring and ensuring process compliance with defined standards, developing safety management systems, identifying/minimising critical safety issues and responding to emergencies. He has successfully led cross-functional teams and delivered training on aviation safety, quality and regulatory compliance for process improvement. Spyros is knowledgeable in ICAO, FAA and EASA documentation and regulations as well as NATO, NBAA, IATA, IOSA and ISO standards.

Conor Nolan

Conor joined Aer Lingus as a Cadet Pilot in 1989. During his flying career to date, he has operated many types of aircraft and instructed in many capacities. His safety career started with emergency response planning and then accident investigation. He was Director of Safety and Technical for the Irish Air Line Pilots Association for five years before joining the Aer Lingus Air Safety Office in 2008. He is currently Director of Safety and Security, responsible for all aspects of the Safety Management System. Currently, he remains as a line pilot on the A330 aircraft.

Anne-Louise Slack

Anne-Louise has over 30 years of experience as a Health, Safety and Environment (HSE) Professional in a variety of industries including smelting, heavy manufacturing, food manufacturing, construction, electricity distribution, engineering and research, as well as with regulatory authorities. She started her career in occupational health and hygiene and moved into management roles with responsibility for all aspects of HSE. Her current role is with CSIRO as Executive Manager HSE. She adopts a strong collaborative approach to her work with a deep understanding that this is critical to achieving sustainable HSE outcomes. Anne-Louise's qualifications include a Bachelor of Applied Science (Biology) and Master of Public Health (Occupational and Environmental Health).

Derek Stevenson

Derek has been involved in the health and safety space of telecommunications for the past 13 years. After leaving the Parachute Regiment, he entered the telecommunications industry as an engineer until 2003. Being a Chartered Safety Professional, Derek has gained extensive experience across the wireless and fixed-line networks. Derek is an active member of IOSH where he chairs the panel review board and interviews applicants for the chartered level, and he is Vice-Chair of the Telecommunications and Broadcast Committee that represents the members of the industry across 130 countries. Derek has also extended his skill set to become a Chartered

Environmentalist and implemented the recycling of the plastic burden from micro ducting.

Mark Sujan

Mark is the Managing Director at Human Factors Everywhere Ltd based in the southeast of England. He is a Chartered Ergonomist and Human Factors Specialist (C ErgHF) with over 20 years of experience developing, applying and teaching human factors and safety engineering methods. He is a Fellow of the Higher Education Academy (FHEA), which promotes and recognises teaching excellence, a Fellow of the Safety and Reliability Society (FSaRS), a member of the Health Foundation Q community and a member of the Chartered Institute of Ergonomics and Human Factors (MCIEHF). He has co-authored 'Building Safer Healthcare Systems', which forms the basis for the national patient safety syllabus adopted by the NHS.

Chapter 1

System Knowledge: Most of the Times Adequate but Sometimes Insufficient

Aikaterini Karakatsani

Contents

Both my safety stories refer to the part of my aviation career as a flight dispatcher for a commercial airline. Working in commercial aviation is extremely exciting and challenging, especially for flight dispatchers who are like the eyes and brains of the pilots on the ground. During this period, I was responsible for creating and assisting in flight planning. My duties included the consideration of various parameters, including aircraft performance and loading, en-route winds, forecasts for thunderstorm and turbulence, airport conditions and airspace restrictions. We could say it is the ideal job for those who love aircraft but are afraid of heights but definitely for those who want to, at least, improve, if not ensure, the safety of flights. Day and night shifts were different in the period when I was working as a flight dispatcher. The big difference between day and night shifts was that during the latter, I was working alone in the office. Theoretically, after all

the training I had undergone as a flight dispatcher, I was aware of what scenarios to expect and contingency plans to implement.

As passengers, when we hear a pilot's announcement about, for instance,
a route change or a diversion to an alternate airport to avoid adverse
weather conditions, most likely it is a flight dispatcher in the airline operation centre who has assisted the flight crew in making an informed decision.
Flight dispatchers not only plan a flight but also monitor it, and they are prepared to instantly provide almost any necessary information and assistance
to the pilots. Occasionally, this creates an impression or illusion that they
are on the flight deck. In reality, they are thousands of miles away from the
aircraft, inside an office. However, the pilots in command of the aircraft with
support from other flight crew members make the final decision, depending
on the information provided by a flight dispatcher.

Safety and Other Objectives: They Do Not Have to Conflict

A winter night, an MD80 aircraft was operating a scheduled flight from
the Canary Islands to the northeast of the United Kingdom. The weather
conditions were normal and the aircraft departed on time. The flight time
was about four (4) hours. Approximately in the middle of the flight route,
the pilot in command (PIC) contacted the dispatch centre through a selective calling system to declare an abnormal operational situation as he had
received an abnormal engine indication. The aircraft was running out of fuel
mid-air without any apparent cause. During the first critical seconds of the
conversation, my primary concern was a possible 'fuel starvation' or 'fuel
exhaustion' situation. The difference between the two cases is that during
fuel starvation there is fuel, but for some reason, it cannot reach the engine;
it is, therefore, possible in some cases to re-establish the flow of fuel and to
regain engine power. On the other hand, when fuel exhaustion occurs, there
is no fuel remaining to supply the engines, which is the worst-case scenario
of this type of situation.

The PIC predicted that there was a fuel leakage due to a technical defect
causing fuel starvation. The early recognition of the problem by the PIC was
vital to deal with this difficult situation; we had to collaborate towards performing an emergency landing. The next step was to discuss all actions that
could minimise aerodynamic friction and keep the aircraft airborne as long
as possible until it could land at an alternative airport. In cooperation with

the PIC, I collected all necessary information such as aircraft coordinates, flight levels and remaining fuel as well as wind speed and direction. Taking all that data into consideration, I suggested the PIC perform an emergency landing at London Gatwick airport (code LGW). The PIC did not agree with this decision and counter-suggested London Heathrow airport (code LHR) as a better option due to its shorter flight distance from the current airway point of the aircraft. The tone of his voice made me realise his sense of responsibility, along with a sense of urgency under these very critical moments. The PIC had to make a correct and quick decision for the safety of the flight. This moment I felt very stressed. Neither by him nor by the situation, but by myself. The most significant stress in such cases is not coming from the possible outcomes of abnormalities but the pressure we exert on ourselves to be able to cope during the development of an event and the confrontation of the emergency.

I had to properly substantiate my recommendation for landing at LGW with information and data related not only to flight safety but also to the operational impact. In LGW, the airline had ground and fuel service contractors and the aircraft would be offered immediate assistance to avoid significant time deviations and exceeding crew duty hour limits. In LHR, instead, due to the emergency situation, the aircraft would still be allowed to land, but the airport slot restrictions would not allow a departure for at least the next 24 hours. This would not cause a safety impact but an operational disruption. LHR was a hub airport, one of the busiest airports in the world, with global flight connections. Each airport slot grants the airline full use of runways terminals, taxiways, gates and all other airport infrastructure necessary to conduct flights. With LHR operating at almost 100% capacity, it's complicated for an airline to attempt to obtain a slot. Without a slot, it's impossible to operate a flight to or from the specific airport. If the plane landed at LHR, the aircraft would stay on the ground until we would be able to obtain a slot for departure. This would cause a significant flight delay, unavailability of the crew due to exceedance of maximum duty hours, compensation to passengers and effect on the aircraft's next schedule. At this point, I hope the reader understands that safety cannot be approached in isolation from the whole operational envelope of an organisation; instead, safety must align with all business objectives and be managed in conjunction and harmony with them.

Of course, the PIC was in charge of making the final decision based on the information provided from my side, but I had to ensure he would consider the operational parameters as well. Finally, the PIC consented and performed a safe landing at LGW. The duration of our discussion was approximately four (4) minutes. After the end of the communication and before the aircraft landed in LGW, I contacted all relevant parties in LGW (i.e. the local airport authority and the ground service and fuel service providers) and made all necessary arrangements for a quick turnaround. I created and filed a new flight plan from LGW to the final destination where the aircraft was based, and the airline had 24-hour engineering assistance. The PIC called me from LGW and thanked for the assistance and quick response while confirming the accurate information about the availability of all services, which were provided at the airport without any problem or delay. From my side, I thanked him as well and emphasised the point that the early recognition of the problem was the key to deal with this challenging situation and to make a quick decision.

However, during my conversation with the PIC, I could not skip clarifying that my persistence was not an indication of disrespect but a result of a high sense of responsibility relating to my duties and willingness to resolve the situation in the most effective way and least possible negative implications. The PIC completely understood my positions. The only difference was that at that critical moment, he considered the safety parameters as a priority. During an emergency, safety comes first. Flight crews want safe flights as everyone performing or using a service within safety-critical environments. Notably, obtaining a slot at an airport is part of planning a flight and not part of managing an emergency. However, when time allows and depending on the nature of the emergency, overall operational parameters can be examined and reconciled with safety objectives. This, indeed, does not mean that safety is not a priority; it means that when the circumstances allow, any options to deal with an emergency must be explored holistically. My perspective is that safety and other objectives of organisations always go in tandem unless safety is threatened imminently and conflicts severely with the achievement of other desired outcomes. In the latter case, safety must and should be positioned first in the priority list.

By evaluating this case retrospectively, I concluded that the need to make a quick decision requires prior and updated knowledge of the technical and operational factors while taking into account the personal perspectives of the professionals involved. The PIC had the knowledge of how much fuel was being consumed. Many variables can influence the fuel flow, such as flying at different flight levels to those initially planned. If these variables had not been considered, the pilot's awareness of the remaining usable fuel might have been incomplete, even inexistent. Maintaining fuel supply to the engines during flight relies on the pilot's knowledge of the aircraft's fuel supply system and reduces the probability of fuel starvation, especially during a critical phase of the flight. From my side as a flight dispatcher, I had the knowledge and expertise from the operational point of view, which was adjusted with the information, knowledge and expertise of the PIC. This combination should result in an effective solution, and so it did. Apart from the weather conditions, the aircraft's remaining fuel and the shortest route airport, the effective and quick aircraft ground servicing was also an important parameter that had to be considered. Even though I knew that the PIC is in charge to make the final decision, I needed to ensure that he is aware of all aspects and options. That moment I felt as if I were in the cockpit with him, managing the same anxiety and sharing the same feeling. The feeling that the right decision should be taken for the safety of the passengers and the crew and the minimum disturbance of flight operations. Teamwork and shared awareness prevailed in this case without the tiniest sight of complacency.

When wondering whether there were alternative choices and decisions to manage this incident, I conclude affirmatively. The pilot in command holds the ultimate responsibility for a flight from all aspects. He could have decided to land at the other alternative airport as the first option considered without contacting the dispatch centre. Instead, he requested assistance and/or a piece of advice. In case he made his decision without consulting with the dispatch centre, we would have to face the operational effects, but there would not be any safety impact, and this is what eventually matters. I would respect this decision and try to assist the best way I could even, as explained above, landing at LHR would cause a lot of adverse cascade effects. Operational disruptions may occur anytime for various reasons and must be managed in an effective way. When it comes to a safety impact, the priorities automatically change and the best decision is the one which will reduce the safety risk impact.

Nevertheless, I appreciated the fact that a very experienced captain counted on my assistance as a flight dispatcher. As per our conversation after the event, I highly valued his position that safety needs to be managed and its balance with other aspects of operations does not occur by chance. From my point of view, I would follow exactly the same rail of thought under the specific circumstances, even though many years have passed from this event. In cases where an unfortunate event occurs, the first step is to detect the proximal cause and remedy it before it passes through different layers and leads to a serious incident or accident. This is what we achieved through teamwork and effective communication without harming the rest of the operational aspects.

If I close my eyes and travel back in time, I see myself through a mirror. On the one side of the mirror, it is me, a flight dispatcher with only a few years of experience, alone in the office struggling with anxiety, having to develop and recommend a plan in a very tight timeline and demonstrating a high sense of responsibility and teamwork. On the other side of the mirror is the PIC, a captain with many years of experience and responsible for an aircraft, passengers and crew facing an unpleasant situation and asking for my assistance. When the aircraft returned to its base, we were smiling at each other and showing gratitude to each other. This is something I am proud of and I will never forget. That winter night, I realised that sometimes it is not enough to just know what to do but how fast you are going to do it. You must occasionally overcome all odds and just do it by combing knowledge with quick decision making.

Missing the Fine but Important Details

During my career as a flight dispatcher, there were several moments that Murphy's Law was coming into my thoughts. Usually, trouble was coming from several directions at once and most of the time during the night shifts when I was alone in the office. Just after midnight, an aircraft (MD80) was operating a scheduled flight from a Scandinavian country to North Iraq. The aircraft departed with an hour delay due to weight limitations and adverse weather conditions. The flight duration was 4.5 hours. The planning and preparation of this flight included several overflight permissions and a demanding flight plan, which had to consider the unfavourable weather conditions and the necessary fuel quantity to perform a direct flight with a fully loaded aircraft. Approximately one (1) hour before the estimated time of arrival, the PIC contacted the dispatch centre through a selective calling system to declare an abnormal operational situation. The aircraft had lost communication for the last three (3) minutes with Air Traffic Control (ATC). Aeroplanes communicate with the ground at fixed radio frequencies, which the pilots change manually as the aircraft moves from one air traffic control zone to another so that to receive instructions from the respective ATC services of the region. This system, for some unexpected reason, did not work.

According to the European Organisation for the Safety of Air Navigation,[1] commonly known as EUROCONTROL, loss of communication between aircraft and ATC may occur for a variety of reasons, some of them being technical and others related to degraded human performance and job-related aspects of pilots and air traffic controllers (e.g. high workload). Aircraft loss of communication with ATC usually happens in one of the following circumstances: malfunction of communications equipment, mismanagement of communications equipment and radio interference. Moreover, someone might not be aware of whether any loss of communication is transitory or prolonged. Either way, such an event has obvious importance for flight safety.

Also, losses of communication can differ significantly in length; it is, nonetheless, those with an impact on daily ATC operations, which have drawn attention to the deviations and malfunctions and led to research and studies about their resolution. Communication losses affect all aviation segments. The phenomenon is not confined to a few aircraft operators or radio communication types. In the thousand reported events which have disrupted ATC since 1999, more than 300 airlines, 12 radio types, 180 sectors and 190 channel frequencies are represented. One of EUROCONTROL's main priorities is to progress with the deployment and usage of suitable equipment in aeronautical navigation service providers, airspace users' offices and military reporting centres to gather from all relevant parties the necessary incident information to enable understanding and reduction of loss of communication occurrences.

Back to my story, all possible causes were discussed with the PIC; radio interference was removed from the equation and mismanagement of equipment was examined, but no deficiencies were detected. Last but not least, equipment malfunction was not possible to identify and no alerts or messages suggested any relevant problem. It was a very stressful moment considering all the possible effects since the pilot was unable to pass and receive useful information to the ATC. This situation could be interpreted as a security threat and result in a military interception.[2] Having calculated the exact aircraft's take-off time and flight plan parameters as well as the latest weather condition forecast over the specific area, I tried to identify the aircraft's airway and waypoint. Simultaneously, I made contact calls to the ATC centre in the aircraft's vicinity and informed them accordingly about the situation.

The fact that there was nothing more I could do to manage this situation and ensure the safety of the flight was one of my biggest fears as an aviation professional, and I viewed it as a personal 'failure' despite not contributing to the problem directly, however unintentionally I had played a role in this, as I explain below. Nonetheless, although according to Murphy's

Law 'if something can go wrong, it will', on the positive side of this law, 'if something good can happen, it can definitely happen'; indeed, it happened. The communication with the ATC was restored after approximately four to five (4–5) minutes. There was no assumption that the loss of contact with the ATC meant that the aircraft was doomed to crash. However, a longer duration of loss of communication could increase the risk of an undesirable event.

By evaluating this case, I concluded that the aircraft had lost communication with the ground for almost eight (8) minutes in total and it was not possible for both sides (PIC and myself) to manage this incident directly and effectively. Various questions started passing through my mind. Why did this happen? Which were the contributing factors? Was it frequency congestion? Was it a pilot error due to workload? Was it an equipment malfunction? None of these questions could be answered. At that moment what really mattered was that the flight was safe and we were back to normal operations. All questions have an answer and all problems have a solution. How can someone deal with multiple questions and problems at once and provide answers and implement solutions in parallel is a different story to narrate. In this case, an investigation was conducted afterwards and concluded that the loss of communication was not the result of frequency congestion, equipment malfunction or pilot error. The PIC implemented all defensive practices. He demonstrated good radio discipline and was aware of alternative processes to follow. As soon as he realised that none of these methods was effective, he contacted the dispatch centre and then it was my turn to act.

After wondering whether there were any alternative choices and decisions to manage this case proactively before the incident occurred, from my perspective, I realised that yes, there were. The most significant omission from my side was that I handled the preparation of this flight the same way as I did with all other flights. However, there was a big difference with this specific flight: the route was not only within European airspace. The destination airport was in Iraq. I should have conducted a thorough assessment beforehand and checked if there were any deviations or deficiencies instead of only securing the overflying permissions out of European airspace. It had been reported by other aircraft there were losses of communication with ATC in this specific airspace zone, but I did not search thoroughly for this information. The truth is that I was not experienced with this kind of flight, even though, by making a strict self-judgment, this may sound like an excuse. It is the same feeling as when you sign a contract with enthusiasm. Since you might be not familiar with contracts, you rarely pay attention to these tiny letters and fine details, a situation that can lead you to regret missing some critical information. My regret was not relieved even if I knew retrospectively that the result would be the same if I had done things differently and had considered additional parameters and the information available. Despite the favourable outcome of this incident, I had missed the opportunity to identify this critical information, get prepared for the possibility of temporary loss of communication between the aircraft and ground and inform the PIC accordingly.

If something can go wrong, it will; at least be prepared for it. The most important is to learn from your errors and not repeat them. The issue I dealt with was discussed with the involved personnel and management and was considered as an additional topic to be examined during the preparation of similar flights. Many years have passed from this event, and I got my lessons. The first step is to prevent an error and/or a lapse from occurring where possible. To achieve that you need to have practical checklists, briefings, maps, charts, manuals and clearly documented standard operating procedures. In case an error, mistake, lapse or just an unfortunate event occurs, the next step is to detect the principal cause and remedy it before it generates ripple effects across your system. This can be achieved through effective communication, teamwork and, last but not least, verification and validation of factual information and data.

Nowadays I do not work as a dispatcher, but I have taken advantage of all this valuable experience from this work, and I always try to be prepared in the best way before I manage a project or/and make a decision by doing research and, if necessary, conducting a risk assessment. The most important thing I have learned as an aviation safety professional is that we must build a proactive management system to be able to develop a predictive management system.

Notes

1. Investigation into Loss of Communication (EUROCONTROL 2008). Retrieved from https://skybrary.aero/bookshelf/books/826.pdf
2. Skybray (n.d.). Loss of communication. Retrieved from https://www.skybrary.aero/index.php/Loss_of_Communication

Chapter 2

Safety Interventions: How Can We Make Them Worth the Effort?

Anne-Louise Slack

Contents

In the following chapter, I reflect on two different approaches I have taken during my Health, Safety and Environment (HSE) career. For some context, I will start with my grounding role with Workplace Health and Safety Queensland (WHSQ) which shaped aspects that were critical to my success stories and something I neglected to do when I failed. When I started working in 1990 for WHSQ, the scope of my role as an advisor/inspector was different from the scope of the role an inspector has today, at least in Queensland. I was young and surrounded by many WHS professionals, each with their specific areas of knowledge and expertise. Collectively there was a massive plethora of experience that I tapped into regularly and never encountered in any other organisation.

This experience set me up for my success stories because I was regularly visiting workplaces in which I had no prior knowledge and needed to take the time to understand what they did and what their risks were. I learnt that I needed to ask questions, listen and provide guidance to employers on how they could improve in a way that incorporated their solutions, while always being in line with legislation. I realised if the employer was part of developing a solution, he/she was more likely to implement it. This role taught me to listen and ask people what they thought of certain risks and then ask how they could best minimise it. It taught me that it was fine to enquire and that I didn't need to know all the answers. In fact, I usually didn't have the best answers; the people doing the work did. For me, this approach has been reinforced in recent years by the work of many scholars such as Sidney Dekker, Todd Conklin and Eric Hollnagel.

People Can Be the Solution: A Collage of Success Stories

My success story relates to my most recent role where I consolidated many years of experience, learning and reflection. It was the smallest organisation I have worked for, and as a research organisation in the agricultural industry, it was potentially one of the most dangerous. There was lots of heavy, often old, machinery, mobile plant and bespoke equipment built for specific research purposes that had inadequate guarding until we upgraded them as much as we could. Add in a considerable amount of manual handling that required a lot of repetition, excessive force, awkward postures and a bit of vibration. Yes, it was a recipe for musculoskeletal injuries, and we had a few. There was a wide range of chemicals being used in the laboratories and on the farms. Many of our people worked long hours in the sun during harvest along the east coast of Queensland and northern New South Wales. One of our biggest risks was the significant number of our people driving on the Bruce Highway. It might seem strange that driving was one of our biggest risks given most people drive every day. However, given the frequency in which our people drove at work, mostly on the highway at 100 km per hour while towing trailers, and the potential consequences if there was an error of judgement, a problem with a trailer or a collision was deemed to be a significant risk. There had been a couple of serious motor vehicle incidents while I was working there including a vehicle that aquaplaned and rolled over into a drain filled with water during inclement weather and another T-bone collision into the driver's side of a vehicle while turning, narrowly missing the driver's door. Fortunately, no-one was killed or seriously injured but the potential was certainly there.

When I started, I was the only HSE professional on staff. A year earlier, one of our senior researchers had two fingers amputated after being crushed in a machine. The organisation was likely in breach of its WHS obligation and faced prosecution by the Regulator. They chose to commit to an Enforceable Undertaking as an alternative to being prosecuted, and this was in its final draft with 22 deliverables when I started. One of the deliverables was to pilot a culture change program because it was perceived that there could be an issue with the culture. I was excited to tackle this issue and knew a consultant who could give us a kick-start. The program was largely psychology 101 viewed with a safety lens and it explained how people could take control of their safety and the safety of fellow staff using several psychological tools. The program also explained that we were moving to a "Safety-II" or "Safety Differently" mindset where we viewed our people as the solution to the problem by using consultation and collaboration, rather than the traditional approach where people were viewed as the problem to be controlled and solved through policies, procedures and a strangling bureaucracy. I worked with the consultant to contextualise the program to the business and co-facilitated the presentations. This gave me critical ownership in taking this mindset forward.

However, we didn't present the first pilot program until 18 months into my role, so I needed to start making inroads into doing safety differently in everything I did. As an HSE professional, I wanted to engage and consult with the people doing the work to bring out this valuable source of information that could improve the health and safety of their work, and more broadly, the organisation. I also had autonomy in the role to formulate my approach and decide how to go about my work. I invested time to visit all our sites as regularly as I could and talk with our people to understand their roles and the work they did. This proved invaluable since I developed relationships with all staff in the organisation, naturally some more than others, and an overall level of understanding, care and trust. Spending time talking, observing and sometimes doing jobs such as carrying bundles of cane and digging for cane grubs gave me great insight into the complexities of the work our people did and how interventions or changes to the job would impact on them. This enabled me to apply a very practical approach to implementing safety improvements, particularly since I had laid the groundwork for genuine consultation and collaboration. I gained immense satisfaction from this aspect of my role.

When people said we needed to amend or write a new Safe Work Procedure (SWPs), I asked if they could do it and explained that I would get it wrong because I was not doing their job and they were in a much better position to get it right. I also asked them to consult with others who did similar work or used the same SWP to ensure a well-rounded, informed and safe way to work. People were more than happy to own their SWPs and did some excellent work, sometimes adding photos to enhance the written word. I still reviewed the SWPs to ensure it made sense to a "layperson", which I was when it came to the details of their work, and that the requirements of the WHS legislation were met. I actively avoided, as much as possible, telling people what to do despite people often expecting me to do so.

Let me describe an example. I was chairing an HSE committee meeting made up of site safety representatives from each of our nine sites via teleconference. Most of the representatives held senior roles at their respective sites, and the meetings were well established prior to me working there. A couple of meetings into commencing my new role, I was asked what I thought of long trousers in the field. A strange question, yes, because wearing long trousers in most field roles/industries is standard, but not in this organisation. When I first visited all the sites, I was quietly shocked that many, if not most, people were wearing shorts in the field. So, when I was asked this question, I flipped it by saying "I have my thoughts, but I'm interested in what everyone else thinks first". The first response was "I refuse to wear long pants; they're too hot". Then another piped up and said that long pants are required when we visit other businesses as part of our work. Several more comments were made either supporting or rejecting the idea. I just listened.

When the conversation started to lull, I chose this time to add my thoughts. "To be honest, I was surprised to see how many people wear shorts here because it is a common and accepted industry practice to wear long pants and long-sleeved shirts to provide protection from UV exposure while working in the sun. We also have the added risks of infection from leptospirosis and melioidosis which are both potentially fatal and enter through breaks in the skin. Our work in cane fields means we get cuts and scratches regularly if the skin is not covered. We also have exposure to snakes, cane knives and mosquitoes that could carry tropical diseases. While trousers are barely protective against snake bites or knives, they do offer an extra layer. I have worked for another well-known company with employees throughout Queensland who all consistently wore long pants and long-sleeved shirts. I have also worked for a smelter that required the same and almost every piece of regular Personal Protective Equipment (PPE) imaginable, and the environment was considerably warmer than the heat in north Queensland". I explained the protective value of long pants and how these usually outweigh the discomfort due to heat. A bit of silence, then one person from our most northern site offered to wear long pants as a trial if I paid for them from my budget. I agreed immediately, extended the invitation to everyone and asked that they pass the word on to anyone else who showed interested. I also said I would brand the trousers with our corporate logo. The intent behind the branding was to create demand for the trousers amongst our workforce. When people saw the corporate logo, they would have known the company would pay for them which is always an added incentive. The orders started rolling in.

An important point to recognise was that people who didn't want to move to long pants were against it because they were so used to wearing shorts. Some of them had been doing this for decades and believed long pants would be considerably hotter than shorts. While long pants are usually warmer, they can also be protective against really high temperatures. They used the heat as a reason not to change. One person felt so strongly about wearing shorts that they said they would resign if "forced" to wear long pants. I figured it was more powerful for people to choose to wear long pants and allow the demand to grow organically than to dictate that "long pants are now mandatory". I could have easily chosen the latter, but I would have created resentment and I had limited means to enforce their use. I was only one person and some of the people resisting the change were managers who would have been responsible for ensuring their teams wore them. I would have become the long pants "Nazi" or "policeman" and could have easily been ignored while I was not present on each of the nine sites. In my mind, that would have been a recipe for professional suicide. I detest the notion of being a safety cop because no-one likes being told what to do. I preferred to take the approach of giving people the reasons why a change may be beneficial and allow them to make an informed decision. Peer group pressure helps as well because as more and more people wore long pants the more it started to become the norm and influenced others to adopt the standard practice by choice.

The transition to long pants occurred over three years and most employees now wear long pants in the field with many being strong advocates. All new employees are provided long pants as part of their PPE kit and they all wear them. Over these three years, we continued to have discussions about wearing long pants and I continued to resist announcing it as mandatory. The main reason was that I could not enforce their use while a small number of our managers were still not wearing them consistently. I hold the view that they are the ones on-site who need to enforce and maintain the standard. At the time of writing this chapter, a more senior manager was planning to have one-on-one discussions with the small number of remaining managers wearing shorts to reinforce that wearing long pants is expected amongst the whole workforce working in the field. It has been a slow but important journey. Interestingly, the person who "refused to wear long pants" voluntarily ordered two pairs.

The only times I provided "instruction" was where we would be in breach of the legislation, and the risk was not apparent or recognised by others. I still do it with a consultative approach, and I am very clear about why it would be a breach of the law. Let me give you another example. We had a hammer mill, which was a bespoke piece of equipment that had been recently guarded. Inside the new guarding, a latch holding the handle of the hammer mill door broke off causing the door and handle to fly away with the force of the machine. Fortunately, the incident only damaged the guarding rather than becoming a projectile and hitting someone. The latch on the door was repaired and a few days later, the exact same incident occurred again because the design was inherently weak. I was told they would repair it again and reinforce the design of the latch once all the samples had been tested after harvest. They had 250 more samples to feed through the machine. A warning flag (i.e. potential conflict with our obligation to ensure plant and equipment was safe) was raised in my head. I contacted the person who reported both incidents to understand what the worst possible outcome could be if the latch failed a third time. She explained the operator was out of the line of fire where the failure happened, but she didn't know if the protective cage had the potential to fail.

I contacted the farm manager for his assessment and again asked him to consider the worst possible outcome. I told him I trusted his judgement and I genuinely did. I knew the machine but wasn't on-site to look at it closely. The farm manager examined the machine with the person who repaired the latch. They both considered that the cage would most likely hold if the latch broke again but decided to weld a steel rod inside the cage in case of another failure. They would still complete the redesign and reinforcement of the latch after the season. You might be asking why they didn't just redesign and reinforce the latch straight away. I asked the same question and the reason was the time pressures they were under as it would take a couple of days to redesign and reinforce the latch and this would render their samples useless. Welding a protective rod inside the cage was a quick and effective solution in case of another failure and prevented any potential harm to people. I was satisfied this met our legislative obligations and the work continued.

Even though it wasn't necessary in the end, I was prepared to stop the work to fix the latch immediately if a serious potential incident could occur with a third failure even though it could take a couple of days and we would lose valuable samples. The safety of our people must always take precedence over the work. I thanked the person who reported this incident. She didn't know if she should report it the first time since there was no injury. She rang again the second time to check if she should report it too because her supervisor said she didn't need to. I thanked her again because it did need to be reported, and I was grateful that she did. Her supervisor was not delinquent or discouraging in saying she didn't need to report the incident; he just thought the original report was enough. What he didn't take time to consider was whether a third failure could cause more significant harm. He and others at the location learnt about the importance of reporting from this series of events.

I have worked consistently to develop a level of trust where people could call me with the assurance that no-one would be penalised for reporting anything. I was making an effort to thank everyone for reporting incidents regardless of what happened, and this has led to a consistent positive reporting culture since I started in January 2017. When a Board member started asking questions about the number of incidents being reported, I explained that this was a positive sign. As an organisation, we needed to know what was happening on the ground. After all, our industry was one of Australia's most dangerous, and our people were still suffering injuries. Therefore, we needed to welcome and be grateful for people reporting incidents, near misses and hazards. More frequent reporting provided the opportunity to ensure corrective actions were followed up and gave us insights into any emerging trends. I gained considerable support from our Board, the CEO and Executive Team as they saw the influence and benefits of this approach combined with other initiatives.

One initiative I introduced to understand the multiple factors contributing to incidents and hazards was the concept of Learning Teams, thanks to the excellent work of Bob Edwards, who is a Human and Organisational Performance coach. He helps organisations realise that human error is common and expected in complex work environments. He demonstrates how, when something bad happens, our focus needs to shift from blaming those doing the work to a look at the system they are working in. I started to replace traditional investigations of significant incidents with Learning Teams, and the feedback was extremely positive. A comment I received from a participant in the first learning team I ran was "I thought I was going to get blamed, but this has been really good, and we are able to avoid similar incidents in future". These types of comments helped to confirm that Learning Teams are a positive way forward. However, more time and education about the mindset behind the Learning Teams is needed to embed the approach across the whole organisation.

So why is this approach so useful in my opinion? Learning Teams are a powerful method to gain greater insights and understandings of everyday operations, whether an incident has occurred, or performance is not efficient or effective. Learning Teams are about improving performance and the key principles are Error is normal, Blame fixes nothing, Systems (context) drives behaviours, Learning is vital, Response matters. Learning Teams move the focus from error and blame to a more holistic, collaborative and systems thinking approach to failure. Emphasis is placed on the importance of operational learning and changing from a reactionary mode to more of a "learn and improve" approach. This is done by bringing workers and planners together to discuss how work really gets done, not how management thinks it gets done. Conversations in Learning Teams become more open and honest as we change our response to failure and therefore get a clearer picture of just how difficult it can be for workers to get the work done safely and successfully. Learning Teams are about our conversations, the way we treat each other, about listening and learning from those who do the work and appreciating the depth of knowledge in the way work is actually done. It's about taking the time to respond thoughtfully. Learning Teams look at the complexity of failure and consider the possibility that it happens more from normal variability than from some anomaly or someone not paying attention.

Now, let me circle back to the culture program and how this was key in changing the mindset to a Safety-II or Safety Differently approach. The consultant and I interviewed a random selection of people from every site to understand what they thought about safety, what their frustrations were and how things could be improved. By doing this, we gained a deeper insight into the culture of our organisation. The consultant also interviewed each member of the Executive Team for their perspectives. We tailored the program around these insights to reflect the broad experiences of people in the business. I rewrote a safety pledge that was developed prior to me with a series of questions that mimicked a risk assessment. I also added a tag line that represented a common theme that emerged while interviewing people across the organisation. To me, it was apparent that people cared about their safety and the safety of others as well as research and the assets we used to carry out the research. "Just Care" was the tag line and a call to action. We piloted the "Just Care" program at two sites as well as our Executive Team and their direct reports. Everyone who participated consistently endorsed rolling it out to the whole organisation. Everyone who attended also signed a poster-sized copy of the pledge after I explained how it came about. Following the success of the pilot program, I received an endorsement to roll it out to the whole organisation.

At the time of writing this chapter, the Safety-II program has been delivered to all sites and most people have embraced the concepts and have a clear understanding of it. Some more programs must be run to catch people who couldn't attend the first rounds. Nonetheless, like any program or initiative, not everyone gets the concepts of Safety-II/Safety Differently when first exposed, and we have encountered challenges with some people transitioning to a Safety-II mindset. For example, after the Executive pilot course, I had a few people call me concerned about how one senior manager walked around site doing an inspection and started to tell people what to do while giving them checklists from his former organisation expecting our people to use them. This is quite the opposite of the mindset of engaging people to find solutions to their issues. Fortunately, this type of activity was not the norm and further learning, examples and understanding are required. I have a strategy to continue to embed and reinforce the Safety-II mindset and I'm grateful most people understood and embraced the concepts immediately. So, we have a way to go but the journey will continue, and I'm convinced we are on the right path. Many staff have expressed their appreciation and support for my work in this space.

Missed Opportunities: Failure to Connect with Others

Another business I worked for was a major Queensland employer. I was one of the four HSE managers reporting to the HSE general manager. We each had our area of responsibility and a team to assist. I was accountable for occupational health and injury management. On entering the organisation, the general manager had two projects he wanted me to roll-out as part of my role. He had implemented these programs in another organisation and expectations were that they would be received well in this organisation. One program was a generic health and wellbeing program that provided education and guidance about diet and exercise for individuals to improve their health and wellbeing. The second program aimed to encourage people to take personal responsibility for their safety. Both programs had excellent content and were enjoyed by many across the business, but we struggled to embed the principles or see the impact on everyday work. This was evidenced by follow-up surveys from the first program that showed not many people had made sustainable changes to their diet or exercise regimes. The second program, in which it was hoped would contribute to a reduction in incident rates, failed to materialise.

Both programs were "off the shelf" with little or no tailoring to the issues, needs or context of the organisation. I think this may have contributed to the initial enjoyment of the programs, but not with any lasting impact. Furthermore, there was no business strategy to embed or support the learnings moving forward. It was left to the individual to follow through. There were several other programs I rolled out to the organisation based on what was considered good practice. They included skin cancer checks, quit smoking programs, flu vaccinations, a health and wellbeing program that measured blood pressure, cholesterol, blood glucose levels and so on for employees as well as executive health assessments. I also implemented a drug and alcohol policy and procedures, including various testing regimes, and I procured and managed an Employee Assistance Program for all employees. I wrote and reviewed various documents to support the management of health and safety in the business such as the Asbestos Management Plan and participated in many meetings to attempt to improve health and safety in the business.

All this work looked great on my resume, and there were benefits to people and the organisation for demonstrating it was committed to the health and safety of staff. However, when I reflected on the role after I left, I felt like I failed. Why? Largely because I neglected to connect with the workforce. While I was in the role, I felt like I should have been instinctively familiar with all the work they did, the associated risks and have appropriate solutions to help minimise the potential for harm. But I didn't, and I later realised that my belief I should be able to do this was unrealistic. Interestingly, early in my career, I was keen to learn as much as I could, and I asked many questions in the process. I can't explain why, but somewhere along the road, I lost my confidence to ask questions and felt like I should just know details of work that no-one could ever be across completely. It was during this part of my career that I felt like I failed, and it became part of my "mid-career crisis" where I questioned if I was good enough and could make a difference.

On reflection, I really limited the impact I could have made in this organisation. I formed the (unfounded) opinion that the organisation "expected" certain things, and I responded in a way that I thought best, but I never took the time to verify with the people that mattered. These were the people doing the work at the pointy end where the significant risks were a daily encounter. Ultimately, I believe this failure to connect with the workforce led to me being offered voluntary redundancy. My role was being restructured and expanded which was a reason to "spill" the position whereby I could take redundancy or reapply in competition with any other interested applicants. I took the redundancy. I didn't think I was adding much value by staying in a "safe" space in my office for much of the time, largely only working closely with my immediate team, peers and manager. It was my choice to work that way, but I realise now it was not the best approach.

Would I have done things differently with the benefit of hindsight? Absolutely! I would have made regular visits to the field to talk with our people, observe the work they did, understand how and why they adopted their way of working, ask lots of questions about their work and their frustrations and how they could minimise the risks they have faced. I would also enquire about the programs we rolled out, particularly what worked well, what didn't and what other programs or initiatives they might like to see and how we could further improve. Another important aspect I missed was connecting more frequently with the various business leaders to listening to their insights, understand their challenges and asking what might help them. What I didn't recognise at the time was that I had the flexibility to get out and interact with the people doing the work and the people leading the business rather than feeling tied to immediate desk-based tasks that needed to be done. I regret not going out and talking with the staff more. It would have made a world of difference to my understanding of the issues in the field, how I would have responded and the work that I focussed on.

There was one success story that I am proud of in this organisation, even though it was largely due to the diligent work of one of my team members and her decision to use a participatory approach with the teams doing the work. I supported this initiative and contributed some thoughts, but the success really came from the consultative approach, something that I had largely, regrettably neglected to do as I had done earlier in my career. Today, and I hope for the remainder of my career, I will always ask questions to understand the people, their work and the organisation and be happy to ask "the dumb questions". In summary, my key tips for those wanting to embark on an HSE career or improve how to lead HSE in an organisation are:

■ HSE will continue to evolve and ways of doing things will change. Stay informed, examine, analyse and embrace new opportunities.

■ Be open, be curious, be respectful and connect with the people who do the work and those that influence the organisation. Understanding their context and challenges is critical to be an effective HSE professional who is welcomed, trusted and appreciated.

■ Regardless of whether you like a person and their views or not, they each have something worth considering. Take the time to listen.

My efforts are focussed on making the job safe with minimal risk. While I don't want anyone getting hurt, I can live with cuts and scratches and maybe even a minor fracture, since they will heal. However, I can't live with the idea that someone could be killed or permanently disabled, including life-threatening illnesses, and I do my best to identify and prevent these possibilities. In the process, I'm mindful of aligning with the legislation or the principles of the legislation if there isn't any specific regulation. When there is contention or politics, I find that focussing on and constantly asking what will keep people safe inevitably helps me steer my way through.

Chapter 3

Only a Few Seconds to Change the Course of an Event

Athanasios Galanis

Contents

The following incidents took place during the period of my service as an Air Traffic Control (ATC) supervisor at an international airport serving both civilian and military aircraft. I was liable for managing ATC staff and leveraging all available resources to ensure the smooth functioning of ATC services and that the airspace and air traffic under my responsibility were safe, smooth and expeditious. Also, I was responsible for resolving problems and reporting unresolved issues affecting the ATC services to the administration of the Unit.

The details below are relevant to the staff involved and are intended to help you understand the role they played in each story that follows. In my stories, the involved personnel were me in the role of ATC supervisor, the aerodrome air traffic controller (AD controller), the ground air traffic controller (GND controller), the driver of the sweeping vehicle and the electrician.

Both the AD controller and the GND controller report to the ATC supervisor. Their principal responsibility is to control traffic within their assigned space; the former controls the movement of aircraft in the airspace and the latter controls all movements (e.g., aircraft, vehicles) on the ground, especially from and to runways and taxiways. Additional tasks of these persons, depending on the assigned space and role, include (1) visual observation, as far as practicable, of operations within the aerodrome, including aircraft, vehicles and personnel present on the manoeuvring area and carrying out airfield and facility inspections; (2) offering of all possible assistance to flight crews in emergency or distress; (3) provision to aircrews with meteorological and other information required for the safe and efficient conduct of flights; and (4) maintenance of awareness of all messages exchanged through the assigned communication channels.

The sweeper vehicle driver is responsible for the cleanliness of the aircraft movement areas according to respective controllers' instructions. This is a typical procedure to ensure that there is no Foreign Object Debris (FOD) on the runway which the aircraft engines could suck with negative implications (e.g., damage and/or a safety-critical situation). The driver must inspect the aircraft movement areas at regular intervals of no more than two hours. If there is a problem with the cleanliness of these areas, then drivers inform the ATC supervisor and clean the area immediately. Until possibly affected areas are cleaned by the sweeper vehicle drivers, no aircraft is allowed to taxi through these areas and all ground traffic is diverted through alternative routes. After the cleaning of the areas is completed, the drivers inform the ATC supervisor to revise any taxi and movement restrictions.

Electricians are responsible for the control, maintenance and repair of all lighting damage across all airport areas. These include the runways, the taxiways, the intersections between runway and taxiways, the space between the taxiways and the apron and all other lights that assist during the approach and depart procedures of aircraft.

You Will Never Know Exactly, but You Can Always Learn

The incident described below occurred while we were on the third and final day of evaluating all aerodrome personnel. This assessment provided various scenarios such as aircraft emergencies, communication problems, an abrupt change in weather conditions, removal of aircraft from the runway after an accident and so on. The assessment was at a group level as well as for each member of the group individually in times of high demand. It was a very difficult and demanding day for all employees and especially for the ATC staff who were tasked to manage both military and civilian flights. Also, the staff were already quite tired from the heavy workload of previous days. The air traffic at the airport that particular day was denser than usual; especially military traffic was 50% higher than a standard day. To cope with the additional task load, the staffing of each shift increased too; while in a typical day the shifts are staffed with five controllers and a supervisor, on that day, the controllers increased by two, that is, eight staff per shift.

I want to clarify that all communications between ATC personnel and aircrews are carried out via air-to-ground frequency, whereas the communications between ATC personnel and the drivers are performed via ground-to-ground frequency. Consequently, the use of different frequencies does not allow flight crews to hear the communications between drivers and ATC staff and vice versa. The use of a different communication frequency for ATC personnel-sweeper drivers and ATC personnel-aircrews does not allow a complete and optimal situation awareness of the dynamic airport environment. Of course, someone would ask why there is not a common communication channel. The reality is that the amount of information exchanged per unit of time among ATC staff, aircrews and ground personnel is vast, and the exposure of everyone to everything that is going around on the air and on the ground could distract individuals from their core tasks, even safety-critical ones. Therefore, it is a trade-off situation that imposes on ATC personnel the role of mediator and holds them responsible for transferring critical messages acutely between different job roles and ensures that the people involved share the same information.

On the day of the event, the runway was used for take-off and landing with the direction of the aircraft being east to west. It is important to note for the incident described, the length of the runway is 11,000 ft; this distance is quite long and, as it will be understood from the incident described below, it contributed positively to prevent the worst-case scenario from becoming real. It was just before sunset and the traffic involved was a helicopter, which was number one for departure, and a formation of four military aircraft commissioned for a training flight as number two for departure. The helicopter was standing in the middle of the runway and was ready for departure, and the military aircraft were standing at the beginning of the runway, also ready for departure. All communications between ATC and the helicopter and the aircraft had run smoothly and all parties had understood the take-off order and had acknowledged their positions. Nothing to worry about in this phase; I only had to monitor the situation continuously to deal with anything unexpected.

The sweeping vehicle was behind the helicopter waiting for its departure to enter the runway and perform precautionary cleaning. As foreseen by the procedures, I informed the AD controller before the helicopter's departure about the necessity of the runway inspection after the helicopter's take-off and the anticipated take-off delay of the military aircraft formation. The AD controller acknowledged the reception of this request and informed the driver for the need to proceed with the cleaning of the runway pending permission from me. Furthermore, the AD controller informed the GND controller about the forthcoming activity in the runway and reminded him of his obligation to set the runway take-off and landing prohibition sign on upon notification that the activity commenced; this means that no aircraft would be allowed to use the runway during the vehicle's movement in the area. The GND controller confirmed the reception of this information. Therefore, the whole "system" was adequately aware through assertive bi-directional communication about the upcoming activity on the runway. Nothing should go wrong.

After two minutes from the communications described above, the helicopter received clearance for departure from the AD controller. Afterwards, I authorised the driver to enter the runway for inspection and cleaning and informed the AD and GND controllers that the runway was occupied. More specifically, I instructed the vehicle driver to enter the runway and inspect only the area where the helicopter took-off. I believe that the information I provided to the driver was clear and explicit enough. I aimed to minimise the inspection time and avoid a significant delay to the departure of the military aircraft, especially considering the increased load of the ATC services that day. After I granted the entrance permission to the driver, I informed both controllers that the vehicle would move in the runway for a while; this was a reasonable statement to make considering the clarity of my instructions to the driver as mentioned above. I received a confirmation signal from both the controllers, so everything seemed to proceed fine and adhere to the protocols. Surprisingly though, this was not the case.

First, despite my instructions, the driver, after entering the runway, started moving along its whole length from east to west but never stated this in the radio. Second, none of the three ATC-related actions that should be executed at that time actually happened. The GND controller did not place the signal to inhibit the use of the runway by the aircraft, the AD controller did not confirm the placement of the signal and I did not monitor the execution of these critical tasks as part of my supervisory duties. The holes of the widely known Swiss cheese[1] started getting aligned and someone had to stop this sequence. But who could this person be when all the actors are the holes themselves? You need something external to trigger recovery actions!

After four minutes, the military formation leader requested clearance for departure. The AD controller granted clearance after he self-confirmed that the runway was free of obstacles; however, the vehicle was still on the runway and the AD controller had not noticed it. The free-of-obstacles check is the most crucial element in the departure clearance to avoid an accident on the ground during aircraft taking-off or landing. Even today, we are not sure why the AD controller did not observe anything, although the sweeper was still on the runway. Someone could assign any cause and place a judgment to this (e.g., carelessness, complacency, high workload, low light due to the sunset) but even the AD controller himself was not in place to explain it. Since the runway sign to prohibit take-offs and landings had never been placed, everything seemed normal to the AD controller and the aircraft formation. The latter could take-off.

While the first military aircraft was in a take-off speed, I inspected the runway visually and realised the safety-critical situation. Here, someone would think that the wisest decision would be to request the pilot to cancel the take-off, but I decided otherwise. Considering the sufficient distance between the aircraft and the sweeping vehicle (i.e., the position of the vehicle and the likely location that the aircraft would leave the ground), the aircraft's speed and configuration (i.e., maximum possible weight due to external loads such as fuel tanks and training weapons) and the runway's length, I assessed that the aircraft could take-off without colliding with the vehicle. Furthermore, if I had instructed the cancellation of the take-off, I might have actually caused an aircraft-vehicle collision (e.g., time for the pilot to process the information, decide and react in conjunction with the momentum of the aircraft) or damage to the aircraft (e.g., hot brakes and fire, possible excursion on the side or at the end of the runway).

Therefore, instead of cancelling the take-off, I immediately instructed the AD controller to continue the take-off procedure of the aircraft and I used the radio to inform the vehicle driver to vacate the runway immediately. Then, I instructed the AD controller to inform the pilot of the first aircraft, who had already seen the vehicle in the runway, that he would continue the take-off on the normal course. The aircraft formation leader, who happened to be the pilot of the first aircraft taking-off, agreed with my instructions, and after a rapid evaluation of the situation, we informed the rest of the formation members to proceed with their take-offs as planned. Additionally, the first pilot informed the AD controller through the radio that had sight of the vehicle and the situation was under control. The time between the realisation of the imminent accident and the final instructions described above was less than ten seconds.

Was it a rational decision of mine? Well, I did not have time to perform calculations and conduct any formal risk assessment that could lead to the optimum mitigation measures by considering all possible outcomes and options and consult with my colleagues. At that moment, it was a natural decision based on my experiences alone, which, retrospectively, I was able to explain above. This, of course, does not change the rationality or naturalism of my decision; isn't this though what we do when reflecting on past deeds and decisions? We move forth and back along the timeline of clues, events and decisions to detect strong and weak points based on rationality (i.e., expected behaviours) and retrofitting our knowledge base. This, in turn, helps us make more successful naturalistic decisions in the future under time-critical conditions.

However, the story becomes even more interesting. What made everyone on the ATC tower freeze during the unfolding of this situation was the driver's reaction after seeing the aircraft taking off and receiving the request to vacate the runway immediately; contrary to the instructions, the driver continued moving towards the aircraft. After the event, the driver attributed his reaction to the terror he felt by the sight of the approaching aircraft and the loud and high-pitch voice of mine on the radio. Isn't this reaction of a licenced and experienced employee strange in hindsight? Why would someone get closer to the threat instead of running away the fastest possible through the shortest route? This felt natural to the driver, as my decision to continue with the take-offs felt natural to me. Could this have happened to me? Could I have frozen too or asked from the pilot to cancel the take-off, thus, unintentionally, instructing the aircraft and the vehicle users get closer to each other and becoming a more imminent threat for each other? Were we to judge the decisions and reactions of others or reflect and learn from everyone involved? We opted for the latter, gratefully.

After the aircraft's departure, I informed the airport manager. I advised all AD controllers to try maintaining calm to be able to continue with their duties during this quite heavy-loaded day. Nevertheless, I also asked the controllers whether they felt fit to continue carrying out their tasks and I received only positive responses. I want to believe that this meant that they had no concern about continuing their work despite the incident, and they were in a good psychological state. I hope that the responses were not the ones we name "socially desirable" due to the reluctance of my colleagues to load the situation further and disrupt operations. At least, no other incident occurred that day and the following ones.

After the landing of the aircraft formation, we asked the leader to contribute to the investigation of this incident and express his views and understanding as long as the event was "fresh" in his mind. During this first de-briefing, the incident was just described by citing factual information. The second time we met, we reported the mistakes made and everyone was given the opportunity to express their opinions on the incident. It is outside of my intention to report here the mistakes of the other actors of this story and describe the reasons behind them; this story is mainly about my role. The most important is that, after analysing all information collected, a note was drawn up and distributed to all flight-related personnel. Moreover, a series of lectures were delivered to analyse the incident and remind the procedures to be followed by the staff to ensure that all flights were conducted safely.

If I were to go back in time, would I make the same decision? The only sure thing is that we didn't have an accident. Was it my knowledge and experience or pure luck? Could the same exact decision under slightly different circumstances lead to an accident, and could what seems now as a success become a catastrophe? I am not 100% sure yet. The only thing I can ensure the reader is that from that moment until today I feel weird when I am thinking about the specific event. On the one hand, it feels uncomfortable that I was a hole in the Swiss Cheese Model,[i] and, on the other hand, I feel comfort when thinking that the same hole, that is me, saved the day. Which label should I accept finally? A hole or an asset?

Any decision cannot be viewed as a given fact even under identical conditions and might not lead to the desired result. It may be affected by the time available, the timepoint it was required, the person who made it and plenty of other factors within and outside our control. We want to believe that each of our decisions is a choice out of many possible options to confront a problem, but sometimes this happens so fast that you cannot even recall exactly all the options your mind considered at that time. The choice I made in this instance eventually proved right, but it does not mean that it was the optimum one. I will never know because the past cannot be fully reconstructed. We will never know exactly, but we can still learn.

It Does Not Always Work as Expected

It was a perfect July's morning and everything was going well. The morning shift had rolled around without any problem, and the ATC personnel were not particularly tired despite being in the last half hour of their shift; at least, I had not noticed any visible signs of fatigue. This story took place at the same airport as in the previous story, and the involved personnel were me, in the role of the ATC supervisor, the AD controller, the GND controller and the electrician on duty. The traffic involved was two aircraft, one business jet (BJ) and one medium-range, narrow-body, commercial passenger twin-engine jet (CP).

From the early morning hours, there was a problem with the illumination of the runway lights; in particular, their intensity was constantly at its maximum. This was a severe problem because extreme brightness would create vision issues during aircraft operations at night. Therefore, it had to be solved as soon as possible and not later than sunset. Hence, the electricians were in feverish work outside the aircraft's movement area, at the lighting control centre; up to half an hour before the incident described below, they had to enter the runway for light inspection.

As instructed in the procedures, I informed the ATC staff about the necessity of the runway lights inspection from the electricians and how important it was to solve the problem with their extreme brightness. The ATC personnel confirmed the reception of this request. Furthermore, the AD controller informed the GND controller about the forthcoming activity in the runway and reminded him of his obligation to set the runway take-off and landing prohibition sign on upon notification that the activity commenced. The GND controller too confirmed the reception of this information. All communications between ATC and electricians had run smoothly and all parties had understood the process. Everyone should adhere strictly to the procedures of entry and exit of the runway and being continuously alert to respond promptly to the ATC personnel's instructions to vacate the runway in the event of take-off and landing. As in my previous story above, there was nothing special to worry about; I only had to monitor the situation continuously to cope with anything unexpected.

I authorised the electricians to enter the runway for inspection, and I informed the AD and GND controllers that the runway was occupied. I received confirmation signals from both of them, so everything seemed to proceed right and as prescribed. Once more though, this was not the case. At least in the particular instance, contrary to my previous story, the GND controller placed the signal to inhibit the use of the runway by aircraft, as expected, and the AD controller confirmed the placement of the signal. Nonetheless, I did not monitor the execution of these critical tasks as part of my supervisory duties, and, once more, I became a hole of the Swiss cheese.[1]

The BJ aircraft approached the holding point of the runway and was ready for departure while the electrician's vehicle was still on the runway.

The CP was on its final phase, 12 nautical miles from the airport, and executing the standard landing procedure. The BJ requested clearance for departure. The AD controller granted clearance having checked that the runway was free of obstacles even though the vehicle was in the runway; I will explain the reasons below.

As soon as I heard the clearance for take-off to the aircraft, I realised the mistake. I could not believe my ears. Immediately, and while the aircraft was entering the runway, I told the AD controller that there was a vehicle on the runway and he had to instruct the aircraft to cancel the take-off procedure. To my surprise and horror, the AD controller did not follow my instructions and the aircraft accelerated. Not only did the controller not comply with my instructions, but he looked at me in amazement and replied, "are you trying to fool me?" After some vital seconds had been lost to cancel the take-off of the aircraft safely, I addressed the controller for a second time telling him with assertiveness, loudly and in a high pitch that this was a command. This was the first and only time, to date, I had to deal with this behaviour under a so safety-critical context. Notably, I do not yet have an explanation for the reaction of the AD controller. The specific colleague was experienced with more than 20 years' service in ATC and was confident in his actions. I can only assume that exactly this confidence made him believe in general that his decisions and deeds actions were right. Nonetheless, his reaction at that time was unique and had never been repeated before towards me or others, at least to the best of my knowledge.

Getting back to the story, immediately after my interaction with the controller, I used the radio to inform the vehicle driver to vacate the runway as soon as possible. The vehicle was in the middle of the runway, and the driver's reaction was immediate. Meanwhile, the AD controller finally instructed the BJ to cancel the take-off procedure. To offer a complete understanding of the scene at the particular time point, the BJ's pilot started the cancellation process after he had accelerated the aircraft to 80 knots, the electrician's vehicle was shortly before vacating the runway and the CP was in the final approach, six miles from the beginning of the runway. Gratefully, the CP aircrew executed a go-around following the AD controller's instruction and the BJ cancelled its take-off and vacated the runway in the middle. All the above lasted less than ten seconds; ten vital seconds once more!

As instructed in the protocols describing responses to abnormal situations, the AD controller also asked the aircraft crew if they needed any extra ground support such as fire brigade assistance, ambulances or anything else deemed necessary. After the negative reply from the pilots and the initial relief of ours, the AD controller's gaze met mine; he apologised for not immediately following my instruction and thanked me for helping him avoid the accident. His act of apologising was what drove me to avoid escalating this issue further.

While we were still alerted by the incident and the aircraft was taxiing for the apron, the worst came. The BJ crew reported a high-temperature indication on the braking system and smoke from the nose wheel. Then we sounded the alarm that activated the fire department and ambulances. The aircraft stopped on the taxiway before entering the apron and waited for the rescue vehicles. Thirty seconds after the first report about the smoke from the nose wheel area, the crew stated they feared a fire on the aircraft. The pilot-in-command announced that he set the parking brakes and would follow evacuation procedures; in total, seven persons were on board. We immediately updated the rescuers on the condition of the aircraft and provided wind data and the precise position of the aircraft. The smoke was visible from the control tower despite the 2 km distance between us.

It was a very difficult time, and we were trying to find ways to contribute positively by all means. An ATC colleague who was recording the incident informed us that the persons on board delayed enough to get off the aeroplane; we froze. But we had to continue to support the situation with every possible action. Hence, we updated the rescuers who had already arrived at the scene. At that moment, the tailgate of the aircraft opened, and people began getting out very quickly, the firefighters sprayed foam on the aircraft, and the rescuers helped everyone to walk away from the aircraft. A few minutes later, our rescuers informed us that the fire was completely extinguished and everyone was safe without any injuries. The handler towed the aircraft to a safe place so as not to disturb other traffic. This was a time of total relief and for reflection on the incident.

During the de-briefing, the incident was analysed; referring only to the facts related to me, the AD controller apologised once more for his behaviour. He also informed me that shortly before the time of the incident, he had received a message from his wife about a serious health incident involving a person of his close family environment. This situation caused him to lose his concentration for a while and not build sufficient awareness of the prevailing situation. In a nutshell, the health problem of the family increased his emotional and mental load, this causing him to underperform. Then I told him that I wished he had informed me earlier so that I could have relieved him from his duties. It is very important in all jobs, but even more for safety-critical roles such as ATC, that staff are in good physical and psychological condition. You need a clear mind and a relaxed body along with a positive mood to be able to work safely.

And in closing both my stories, I would like to refer to Human Factors. Was suboptimal human performance a reason that led to these incidents? Yes, it was, but the superior human performance was also the solution! In these stories, it was me who, despite my imperfections (e.g., inadequate monitoring), dealt successfully with the events. But on other occasions, it was someone else who compensated for my mistakes and saved the day. This happens every day, hour and moment to everyone. Human factors deal with opportunities and threats affecting how people perform their jobs; although we often listen about Human Factors after an unpleasant situation, we must bear in mind that this field is also about positives. If we study human factors, we can explore the human body and mind to understand our capabilities and limitations and help ourselves and others. It is not about judging but improving. To my view, the topics of communication and individual and team decision making relate to crucial social and personal skills that complement our technical ones and contribute to the delivery of safe and efficient work.

Note

1. Reason, J. (1990). *Human Error.* New York: Cambridge University Press.

Chapter 4

How Could the Use of Technology Support Safety Management Programmes?

Badar Farooq

Contents

My stories are from my time as a corporate safety advisor at an Engineering, Procurement and Construction (EPC) contractor located in the Middle East. In addition to regular management system related tasks, I worked on several different initiatives designed to enhance the company's safety culture and industry image.

Innovative Safety Solutions? Check In-House Capabilities First!

My employer's corporate safety team was toying with the idea of reinforcing the company's safety rules in the minds of all contractor staff at various construction sites around the world. This was deemed important for two reasons. First, blue-collar workers spent most of their time at the proverbial 'coal face', meaning that they were the ones most exposed to hazards posed by tools, equipment, procedures and processes. Second, many of these workers were not proficient in technical language, so it was considered that a hard-copy visual live training tool would be more effective at conveying relevant safety messages; this seemed to me to be a reasonable initiative. Modern studies, as well as my personal experience, corroborate the fact that audio-visual learning experiences, and especially those that involve an element of 'gamification', are highly effective in engaging participants and imparting knowledge.

The corporate team spent a few months developing A1 size plastic posters depicting the main messages contained in these safety rules; the creation of these posters was a top-down decision coming from the head of the department. However, it was well-received by the team; the other options available were the traditional ones such as rolling-out the safety messages via classroom PowerPoint presentations or verbal toolbox talks at each site. These communications means had been applied already at other instances, and due to the reasons outlined in the previous paragraph, we sought a new way to impart our safety messages effectively. The team did not have previous experience in employing safety posters to disseminate safety messages, so this was meant to be a learning and challenging journey. Initially, the posters were intended to be used by safety trainers at project sites during short stand-down or toolbox talk sessions. During the latter, instead of simply speaking, the trainers could enhance their messages and delivery impact by drawing attention to the depictions on the posters; at the same time, they could test the audience's understanding by asking questions related to the images shown.

After receiving the final posters from the printing company, the critical question about training our site trainers on using the posters appropriately arose among our corporate team. This was especially important because the posters were merely visual, and the team wanted to ensure all trainers and we were on the same page. We were to ensure everyone would impart a consistent message about the company's global safety rules across our own and contractors' employees working in different geographical regions and project teams. Thus, our department had asked the team to ruminate and come up with an innovative and cost-effective idea to roll-out these posters across the different project sites.

A couple of colleagues put forward the usual solution of having corporate team member(s) visit sites where they would sit down with the trainers and explain the posters and their use. This would obviously cost the company time and money as our team members would fly to projects located across the Middle East and Asia over several months. Owing to my personal interest in consuming video content, as well as producing some of my own over social media, I began thinking about a quick and easy way to transfer our knowledge and methodology over to our trainers using a digital medium. Videoconferencing was a potential solution I considered, but it had a couple of drawbacks in our case. Some of our construction sites had rudimentary IT and internet facilities, and thus, holding a videoconference would be challenging, if not impossible. Also, every time a new safety trainer would be recruited to a project, he/she would have to be trained via a videoconference call; hence, corporate office resources would have to repeat the same task at different points of time which would be a hassle considering that corporate employees typically get fully occupied with other important assignments after completing previous ones.

Therefore, I came up with the idea to record voiceover training videos for each poster, explaining in layperson's language how each poster was intended to be used and the messages it should convey to the audience. I discussed the idea with some colleagues, and I received positive reactions. Then I pitched my idea to the head of my department, who welcomed the innovative solution and tasked me to examine its feasibility and plan its execution across our projects. Our company had engaged third-party media agencies in the past to produce short safety videos for various topics such as incident reconstruction and learnings. I first contacted one of these agencies to discuss our requirement, the solution they could provide and respective costs. I was informed that the production of a few simple videos in a single language would cost the company around US$10,000. I intended to have the videos produced in, at least, the three most commonly understood and spoken languages across our project sites – English, Arabic and Hindi/Urdu. Consequently, this was turning out to be an expensive undertaking for what I had considered being a relatively simple exercise.

As a corporate initiative that had to be rolled out at the earliest, especially since the posters were ready, I knew that my head of department would have most probably signed off the requisition to get these videos made via a third-party agency. Nonetheless, I turned my thinking towards producing these videos in-house for a number of reasons. I was driven by lowering the cost of production of course, which was a typical goal of every business department at the company, but also by curiosity and temptation to test my creativity on this initiative, accompanied by the prestige that successful implementation would bring to me and the department. It might sound somewhat ambitious and risky retrospectively, but back then I believed that our team needed something self-made and effective to demonstrate that we were able to create 'safety material' beyond the standard ones (e.g., procedures and training requirements) and we were genuinely interested in promoting safety across the company through more engaging means.

I figured we could use the camera of a high-end mobile phone to do the job and a quiet conference room to record ourselves explaining each poster and how it was meant to be used. I was aware that phones were already being used for recording decent quality videos, especially indoor, and these videos were being shared for mass consumption over the internet. I wanted to extend this option to producing our internal educational videos. I managed to engage a couple of colleagues to help me with arranging and executing this project. We booked one of our conference rooms where we placed a mobile phone in a stable position in front of a white screen. We put each poster in front of the screen, so that the camera was only focusing on it, and we had one of us sit nearby, point to the different areas on the poster and speak in the background to explain its utility.

Since for all of us involved this was the first time recording a 'formal' video, the process took a bit longer than expected. We initially estimated a couple of hours to compile each video, but it took up most of the day. We had to edit the recorded video or even repeat the recording until we had a final video of acceptable quality, meaning that the video was not grainy and the audio could be clearly heard when played over a speaker. As intended, we produced videos for each safety poster in the English, Arabic and Hindi/Urdu languages. The head of department viewed the videos and signalled his approval for their use across all company projects.

Afterwards, we shared the videos with the safety teams across all our project sites from North Africa to South East Asia. The utility of these videos has surpassed my expectations. Other than being viewed only by the safety trainers to educate themselves on the use of the posters on safety rules, during a recent visit at a project site in the Middle East, I noted the videos were being shown by the trainer directly to the final audience (construction workers) in a classroom setting. Upon inquiring with the respective trainer as to why this was happening, I was told that he found the videos were transmitting the message very effectively, perhaps better than he thought he could do himself. He also stated that he could captivate workers' attention in the classroom by employing multimedia for training, as opposed to delivering the message via 'simpler' means in a field toolbox talk holding hard-copy posters. This meant that the corporate office had, in a way, 'killed two birds with one stone', effectively training both our trainers as well as the target construction personnel audience with one innovative and cost-effective solution by utilising in-house resources. Even more importantly, all new company projects to date are supplied with these videos to support educating their workforces on the company's safety rules.

Of course, someone would reasonably wonder whether this initiative delivered in terms of safety performance: had the posters and videos achieved their ultimate goal and contributed to increased safety levels? In our company, we did not specifically track the effects of disseminating information about our safety rules (addressing Permit-To-Work, safe driving, working at height, management of change etc.) on our recorded and reported statistics. The company was running a number of other corporate safety initiatives also centred around these safety rules, such as a safety boot camp for supervisors. The safety rules were also shown in onsite safety inductions to all employees. What I can share after reviewing the company's overall safety statistics over the past three years is that the recordable incidents frequency rate progressively decreased. Attributing this trend only to our in-house safety videos being shown across our projects would be arbitrary and unfair. Safety performance is the result of collective initiatives and efforts through all means and across all organisational levels, and under a systems approach nobody can convincingly argue who and what plays a less or more significant role. What can be of relatively lower effectiveness on one project site can be highly impactful on another, depending on a multitude of factors (e.g., workforce compositions, supervisory capacity, local leadership, working conditions).

Nevertheless, what I can testify is that my drive to mobilise internal resources and employ impactful multimedia educational resources, especially in time-deficient environments, gave rise to an innovative and optimal solution. I certainly risked the department's and my reputation by going down an unfamiliar safety education solution route, but thankfully it worked in our favour. Since we had sufficient time and collegial support to prepare the videos in-house, we were able to mitigate the potential risk of producing shoddy videos. The fact that we did not burn any departmental budget while doing so also helped. Hopefully, my company and others can take inspiration from this effort and come up with similar and new effective solutions to build workforce competencies.

Beating Administrative Burden via Digital Solutions

My employer's management decided to introduce several safety initiatives company-wide a few years ago. The intention was to raise the profile of the safety function across the company, engage clients and contractors and invigorate the safety and construction teams with new tasks and leading targets. These tasks and targets were primarily associated with what the management perceived to be major safety-critical areas in our operations, such as site driving, mental health and inexperienced subcontractor staff at new project sites. One of the initiatives was the Short Service Employee (SSE) program that has been in place in a few companies across different industries for a while. In brief, the SSE program identifies project employees that are relatively new to the site (e.g., in their first six months of employment) and treats them in a special manner, namely:

■ They undergo relatively more frequent coaching on pertinent safety rules related to the site. The reason is that new staff are deemed to be at higher risk compared with employees who have been at the site for longer and are more familiar with local rules and conditions. It should be noted, nonetheless, that employees experienced at a particular site could also be at risk due to holding cognitive biases linked to their familiarity with their work and competence, which may lead to ignoring red safety flags. However, such employees are managed via functional inspection processes and other training programs.
■ They are identified easily at the site via coloured stickers on their hard hat, or a different coloured hard hat, to let the others know that these employees are at a higher risk of harm to themselves or other personnel/assets. Therefore, work crew supervisors pay extra attention to novice employees and do not assign them to hazardous tasks and activities during the initial employment period. I want to clarify that new employees have never expressed any negative feelings associated with this tagging for the SSE program. The tagged employees consider these as part of normal site procedures where their stickers are similar to the other stickers attached to every employee's hard hat, indicating that they have attended the mandatory safety induction.

The impetus for the SSE program was that, according to the analysis of incident data across the company for the past few years, many (contractor) employees who were involved in accidents were found to be new to their particular site. Notably, the company did not analyse the composition of the staff involved in accidents to a detail that would allow any statistical correlation of the figures (e.g., examining the involvement in safety events relative to the percentage of new employees against experienced ones or considering the types and durations of the tasks undertaken by each cohort). However, based upon factual comparative data reported back from all company sites that examined the number of workers involved in incidents versus the time they had spent on the site, there was a clear and negative correlation; employees who had been at the site for a short time had been involved in more incidents. Therefore, we were primarily concerned with workers that were to be involved in day-to-day construction tasks and had arrived on to the project site for the first time. Hierarchical levels from construction supervisors and above, who were managing but not performing the actual tasks, were exempted from this program.

Our head of department tasked a colleague and me to plan and implement this initiative over the following months and enlist the required support of other colleagues and departments. After several brainstorming sessions, we decided on the tools that were to be developed and rolled out to a select number of initial project sites to pilot the program. We worked hard at developing a corporate procedure and program methodology, a data recording spreadsheet and a presentation to be delivered at each site. Out of these, the data recording spreadsheet was the most critical and time-consuming activity as we had to engage specialists from the corporate IT team to customise an Excel file with the addition of macros and background code. The intention was to have a file that would only require entries of SSE employees' basic details, their dates of first induction and subsequent training. The file would then automatically alert the user as to when the next training was due or if it was overdue. In addition to this, the file would also depict some analytic tables and charts to provide a quick snapshot of where that project stood in terms of SSE employees and their training status.

After the preparation of the tools described above and consultation with the head of department, my colleague and I decided to visit and introduce the SSE program at three of our project sites in the Middle East. We visited each of these sites during the following weeks, where we first met with the safety managers and safety trainers. We spent most of our time explaining the program and handing over the materials to the trainers, who would be the program's liaison point and driver onsite. None of the trainers was familiar with any similar program or had implemented such a program previously in their careers. This along with the feedback received during my conversations with a couple of trainers made me a bit apprehensive on the successful implementation of the program.

I got the impression that this could be an additional administrative 'burden' on the trainers, especially on busy worksites with a large regular influx

of contractor staff that were undergoing frequent induction and other training. The trainers were responsible for delivering and documenting all the
training. With the global increase of safety statistical reporting and analysis
along with the recorded data that feeds into it, safety teams being burdened
with administrative data collection and reporting tasks is a common phenomenon. This is at least what I have noticed across all project sites that
I've visited in my career. Our company's corporate safety management was
well aware of this, but on lump-sum lean EPC projects, adding on additional human resources to assist with such tasks is often out of the question.
This is why we, the corporate staff, attempted to streamline and automate
the process as much as possible by using coded spreadsheets. We aimed to
reduce the site trainers' task to only data entry. Nevertheless, we shall not
neglect that busy people on busy project sites are naturally hesitant to add
to their workload, however small the additional task might be; hence, the
trainers considering this as an additional 'burden' was not unexpected.

None of the trainers expressed serious concerns about the SSE program
we had introduced to them; perhaps they did not want to appear uncoop-
erative with the head office team and jeopardise their position in the project and the company. I base this assumption on my experience suggesting
that project contract employees had limited interactions with department
heads sitting in corporate offices elsewhere and the main priority of the
former was to be hired on the next company project. My colleague and I
were ambassadors of this corporate safety initiative which was devised and
mandated by our safety management. Therefore, it was imperative for us to
press ahead with rolling it out as effectively as possible at the earliest possible time. Our department management had already presented the initiative
to the company's executive management as part of the planned safety initiatives, and their expectations on deploying it were translated down to us as
initiative 'owners'.

During the year following the implementation of the SSE program on the
three sites, I followed up with the safety trainers at regular intervals to check
how the program was running. This involved receiving the SSE spreadsheets
from them to gauge how the program was being recorded and implemented.
I was also privy to the number of employees being inducted on each project
since this figure was being reported separately to us as part of our regular
monthly safety statistics received from all projects. I immediately noted the
discrepancy between the number of people reported to be inducted at the
site and those registered in the SSE program as per the spreadsheet. For a
particular site with a large influx of new workers every month, the number

of SSE employees due for refresher training was very large – a couple of hundred employees. Another issue I noted was of a technical nature in the spreadsheet. A project site had somehow tampered with a locked spreadsheet file and managed to change its functionality so that the automatic macros would not work and alert the user to when the next SSE training was due for relevant employees; this actually rendered the spreadsheet ineffective. It turned out that the particular trainer wanted to use the single spreadsheet to record all other training he was also conducting, and unintentionally modifying it damaged its embedded macros. In a couple of conversations with trainers during that time, I was also informed that in some instances, second/third tier small subcontractors would have their employees arrive at the site and begin working, especially for urgent tasks, without undergoing the mandatory safety induction! Obviously, such employees were not recorded in the SSE spreadsheet either and thus posed a potential risk to themselves and their colleagues of being involved in an accident while being unaware of the site safety rules.

My colleague and I noted these shortcomings and went back to the drawing board with IT to update the SSE spreadsheet and make it more robust. We also discussed with project safety managers the issue of appropriately inducting all new employees and appropriately registering them in the SSE program. This was a problem that everyone was cognisant of and trying their best to overcome via engagement of contractor management. Following the results from the pilot program and the rectification of the spreadsheet, we rolled out the SSE program at our remaining projects over the next few months. Currently, it is fully implemented at all project sites under the company's control.

I must note though that the issues mentioned above have not yet been completely eradicated. Whereas employees' behavioural issues and contractor management problems are addressed continuously, through interpersonal communication in many cases, I am particularly concerned about the IT and technical issues surrounding the implementation of similar programs in companies and projects around the world. In my experience, I have found the administrative side of project safety, especially around data recording, analysis and reporting, to be a universal pain point. Data collection and analytics are becoming more important as many companies in various industries around the globe attempt to understand better and, consequently, improve their safety performance. Accurate data informs on a company's safety performance in an objective manner at a particular point in time. Benchmarking this data against previous or future statistics would inform on the performance journey, the effectiveness of any implemented initiatives and so on.

Writing this story in 2019, I am aware of the availability of many mobile applications and cloud-based digital solutions around data recording and analytics, catering to addressing the afore-mentioned pain point. I know that several trailblazing companies with a progressive and innovative safety culture are already employing such solutions. Some have had custom solutions developed to fit their operations best. Such digital solutions prevent exposing safety data to manual human recording and errors and lower the administrative burden on safety teams. Thus, instead of getting bogged down with laborious recording tasks that can be facilitated via technology, safety professionals can invest more of their time on detecting and resolving safety issues.

Chapter 5

How to Eat an Elephant: Implementing Organisational Culture Change

Conor Nolan

Contents

There is a saying that trying to change the safety culture of an organisation is like trying to steer an ocean liner. It needs to be planned, takes time and happens slowly. Really? I recently had the good fortune of becoming acquainted with the captain of an ocean liner, in fact, the skipper of one of the largest passenger cruise ships in the world. Contrary to common understanding, he can stop his ship from full speed in one nautical mile and, in an avoiding-action manoeuvre, can turn it through 90 degrees in about 80 seconds. Wow! Impressive, I thought. But let me tell you, the passengers will know about it, and it won't be pretty; trying to change a safety culture in such a manner might be equally disruptive. My stories are about two attempts to influence safety culture in my company.

Cultivating Future Safety Leaders

My first story is about an on-going effort to influence a particular subset of employees within our organisation: staff applying for promotion to supervisory positions, such as commanders and cabin service leaders, which are all key agents of change in any airline. Getting these people on board with any change agenda dramatically improves the chances of success. Thus, given the opportunity to spend time with every promotion candidate, either in small groups or one-to-one, I saw an opportunity to introduce a change in culture by tiny incremental steps. But for it to be a success, I had to identify what was the benefit of changing to that person, whose focus was on the goal of promotion, not changing the world!!!

The candidates would come to me seeking answers to specific questions: what is my role within the Safety Management System (SMS)? How can I improve my attractiveness as a candidate by demonstrating knowledge of the SMS? Few, if any, had come to me wondering how they could help to change the safety culture from a new position of authority. My challenge is to lead them on a journey of self-discovery where they come to recognise the opportunities to lead, to influence, to change the way things are done, at least within their own sphere of operations. I start by helping the candidates understand that they are already competent and qualified to undertake the role they have applied for. We must accept that our experience and competence will not help us in any way to stand out from the crowd when it comes to leadership positions, as every other applicant has the same skills, training and experience. What can make us stand out is our potential to lead, our ability to influence.

I start by taking the candidates to a place outside of work, helping them identify qualities and experiences from the life that they possess, and which may mark them out as different. That kids' team we coach, the local community committee we chair, the neighbours we care for; these are all life skills that are complementary to being a positive contributor to safety culture. The ability to empathise and try to understand the motivations and influencers of others helps us to recognise people around us who may be able to help us improve our own performance. In many cases, the candidates start wondering where this conversation is going.

Then, I bring them back to work. "When do you put your game face on?" I ask them. In most cases, the immediate answer is when they enter the briefing area, crew room, dispatch office etc. I ask them to reflect on their journey to work, their walk from the car park or transit terminal. "Think about all the people you interacted with along the way, even those you did not notice. How do you think your uniform influenced the way they think about you?" I suggest that every person we meet along our journey to work is a potential teammate; maybe not today, or tomorrow, but someday we may rely on that person to be our critical team member to help resolve an issue or avoid an incident. People say you only get one opportunity to make a first impression, and we all know how long this impression can last.

Therefore, for people in safety-critical roles, from pilots to cabin crew and engineers to ground staff, every meeting, casual greeting etc. should be looked upon as an opportunity to make a connection. That cleaner I meet in the car park might be the person who finds the broken tabletop with the sharp edge; the security agent frisking me might be a pain, but her/his diligence is critical to spotting the malcontent behind me in the line; the gate agent I say hello to might be the one who stops the disruptive passenger from boarding my aircraft. We should never miss an opportunity to build our network. None of us can keep the operation safe by ourselves; we must always rely on our colleagues around us to work as a team. The quality of that team reflects our safety culture.

I arrive at work and I meet the crew. Do I get straight down to work, cut to the chase? Or do I take a few golden moments to make a human connection with that other person or persons in my crew? Humanity is oft forgotten in today's busy world of productivity, time management and high performance. But in a safety-critical operation, the overall mental, physical and emotional wellbeing of our workmates can be a critical element in our success in delivering safety performance. It is not good finding out after take-off or during a difficult maintenance task that my teammate is distracted by personal issues or under the weather.

All of us have busy lives, demanding family duties, hopes and worries. As professionals, we assume that we leave all these back in the car when we lock the door, but in the real world, we are all still mere humans struggling to be the best that we can be, burdened by stressors and distractions. When I form a bond with a colleague, I open the door to dialogue and understanding. Creating an atmosphere of trust within a team, from two pilots in the cockpit to a full shift of loaders, can fundamentally alter how that team performs. Knowing I can share my concerns, knowing I can count on a little understanding, can lead me to try a little harder, focus a little sharper, eager not to betray that small piece of humanity shown to me. As safety leaders, we can all use this technique to bond our teams, have our followers trust us, respect us, and most importantly, do the right thing for us.

So, I've formed my team and reviewed the plan for the day. I have separated the routine from the specific threats on that day and now I must go to work. Planning, communicating, adapting and learning are all key ingredients of resilience. Recognising and respecting those elements that are routine and those that are different today is key to helping prioritise and focus energy where it is really required. A good safety leader respects the training and experience of their team and can quickly process the ordinary, everyday threats. This frees up time and capacity to examine the threats that are different, unusual, challenging perhaps. Of course, in 99% of the cases, none of these unusual threats actually emerge, so planning for the day should be concise, relevant and timely. I don't want to waste time reviewing the mundane. I want to get out there and make more time to solidify the team in the working environment.

I ask the candidates how they get to the workplace from the preparation area. "What a strange question", they reply! Do they walk together as a crew, continuing the process of ice-breaking and team building? Or do they scurry along as individuals, mobile devices clutched in hand, one ear to the phone and one eye to the line at the coffee dock? A good safety leader can turn this short journey into a positive experience, both for the immediate team but also for the wider organisation.

Picture the famous scene of Leonardo Di Caprio walking through the terminal at Miami International, with a crew of flight attendants with arms linked, in the movie "Catch Me if You Can"! Everyone stopped to admire the crew, who were clearly well bonded. It might be only a movie scene, but the sentiment of a tightly bonded crew, laughing and chatting, is a powerful symbol of safety assurance in any organisation. The impact on passengers, some of whom are nervous flyers, is immense. Obviously, the crew aren't worried about the weather or the serviceability of the aircraft if they are so relaxed. We have to remember that those passengers will be a part of our extended crew in just a few minutes. If I need to make a safety passenger announcement during the taxi-out or before an immediate return to land, isn't it better if the passengers have an image of a confident, competent crew? Aren't they more likely to listen and respect me? Similarly, stopping to say "hi" to the gate staff, the dispatcher, the mechanic, makes them feel part of the crew.

Furthermore, Just Culture is a theme that candidates always want to explore and understand better. The policy may be clear, the procedures accessible, but, of course, almost all the practice happens behind closed doors and is not easy to discuss in public. Therefore, I approach it differently and encourage the candidates to reflect on how they can bring Just Culture into their own operation. How do we conduct their briefings to establish clear ground rules, the chain of command and communication protocols? How do we eliminate ambiguity around the role of Standard Operating Procedures (SOPs), especially the ones that get "bent" routinely on the line? What is the role of a safety leader in leading by example? What does that really mean?

There are examples of SOPs where familiarity can breed contempt, and while amongst an experienced and tight crew, this may have zero impact on safety, unless and until all of the crew are that tight, such deviations can cause stress and uncertainty. Without betraying secrets, the cockpit access rules can be a fertile ground for a discussion about SOPs. There is a right way and a casual way. The latter is often fine once trust is established and all crew understand that this is so. But when setting the tone and creating the Just Culture platform, it is a perfect example of how to lead by example. No one should fear following the rules, and a leader that visibly demonstrates that lack of fear, inspires others to so behave.

Dealing with errors is another area that allows us to explore how to bring Just Culture alive. It is too easy to say that I expect my staff always to be professional and that aviation has no room for "B players". However, even the "A team players" can have an off day and make slips and errors that may well be out of character. It is how I collectively deal with these situations that marks out a super team from an ordinary one. Safety leaders will have already created an atmosphere where people know they will not be punished or criticised and that can inspire superior performance. But even if that isn't so clear, how I treat the offenders and how others see me doing this can have a profound influence on the safety culture at that moment.

To say that no one will be punished or belittled is not to say that errors will be easily tolerated, diminished in importance or indeed overlooked. In fact, effective safety leadership demands that errors be promptly and positively dealt with. The person who has made the error or mistake needs to be helped to understand the importance, shown or reminded how to do the task correctly, and then given an appropriate opportunity to reflect on the error and then to discuss what they have learned. If this is done openly and in front of others, everyone will recognise that the person at fault is being treated fairly and with respect, and, if honest, they will realise it could have been them. The overall impression should be very positive as everyone learns, and everyone comes to appreciate what Just Culture really means.

For many candidates, the intensity of the conversation is such that they stop taking notes to add to all the technical notes they thought they needed to be properly prepared for the interview. Now they are starting to really think differently about the role, and the leadership role they can play. I leave them with a challenge; look again at the people you work with, the managers, supervisors and leaders that are already out there. Now reflect on who does all this well and who maybe not so well. Who do you look forward to going to work with when you know it's going to be a fun and fulfilling day and who do you dread? Can they now see that the former actually encourages better performance, and when things do go wrong, it is much more likely that the team will cope and recover because they really are a team? This is what resilience is all about.

Finally, a good safety culture needs to be also a learning culture; thus, safety leaders must understand the role of safety reporting. It is true that everyone is responsible and empowered to report, and in many cases, aviation safety reporting is quite mundane and repetitive. The list of mandatory report items is framed around all the issues we already know about, the "known-knowns" as Donald Rumsfeld once described.[1] But I must make the candidates for leadership roles understand that they have a more important role to play. Through the leadership of teams, we see and hear things that individuals may not pick up on. We must come to recognise activities that are prone to mistakes being made, shortcuts being taken, SOPs getting bent out of shape. Once we comprehend how our inputs to the management system can help to influence the behaviour of others around us, we start to see reporting under a different light, especially when we realise that, in fact, we may directly benefit ourselves from learning about the experiences of our peers.

We indeed try to explain the above to all staff during SMS training sessions etc., but when we play on that natural human instinct, the loneliness of command, the responsibility of being in charge, it is possible to really influence the way these new safety leaders approach their new role. Therefore, by leading a candidate through this conversation about all the things they already do, I manage to make the subject of Safety Management something real, and more importantly, something with direct and personal benefit to the candidate.

I will finish this story with a message which also relates to my next story: changing safety culture is like turning an ocean liner; it needs careful planning and coordination. If we try to do it too quickly, we will suffer damage and fear. Planning and coordination are important, but real success comes when everyone shares the vision and sees the benefits, not just to the organisation but also just as important to ourselves. Changing a safety culture may seem like an enormous task in a big organisation. I hope I have shown here that if we target key individuals within that large group, it is possible to have a disproportionately positive influence. How do we eat an elephant? One bite at a time!

Organisational Inertia When Faced with Innovation

Several years ago, faced with the prospect of imminent changes in regulation, I foresaw a requirement to make changes to our internal safety reporting system. In planning these changes, I also recognised opportunities to improve the system, eradicating legacy practices, streamlining workflows and all the good stuff the comes along with effective change management and project planning. I thought I had it all figured out. Months of working with the software supplier to develop and test the new tool, meeting after meeting to document the new policies and processes, and finally, a view of the horizon when the system would be ready to launch. All I needed now was the support of the management team, my peers and my safety management colleague. This should be a rubber-stamp step in the process; I thought that the management would approve the newborn safety reporting system without much scrutiny.

I brought the safety review board members together and presented the new system to them in what I thought was "a masterpiece of PowerPoint excellence". Background, context, regulatory obligation and pure, brilliant efficiency were explained in simple, inoffensive language. And anyway, we had no choice because the regulation would demand this soon. What could go wrong? I was greeted by stunned silence.

In making the changes to the system, I had eliminated one single step that was a legacy throwback to the days when the world was connected via SITA telex; the latter is a communication tool – basically like an email address, but it's a standalone system that is widespread in the travel industry. For decades, every time a reporter, almost exclusively the captain of the aircraft, needed to submit a safety report, he/she first needed to alert the company that something had happened. This was done by phoning the duty manager in Operations Control and raising an "Incident Signal" which was composed based on a list of required data, then transcribed and sent around the network by Telex. It was a way to alert management that something had occurred, allowing an opportunity to react. In later years, it became an email, which facilitated the number of recipients growing to enormous numbers. Imagine what that prospect meant to a pilot considering self-reporting an error? One can see why this step needed to be eliminated, notwithstanding its redundancy in this age of cell phones, ACARS, Flight Tracking and on-board WIFI. Furthermore, with the blossoming of safety reporting into all areas of the business and the constant growth of the number of flights, you can imagine how "happy" the duty manager was with the phone ringing every five minutes with someone reporting a bird strike, a damaged tow-bar or two bags left behind!

Therefore, I had assumed that the introduction of a modern reporting system with on-line and off-line mobile capability, auto-generated daily reports and all the bells and whistles we have come to expect from an SMS tool, would far outweigh the attraction of a legacy, antiquated messaging system. I was so mistaken! It turned out that most, if not all, of the management team, had become reliant on the simplicity and timeliness of the Incident Signal, which by now arrived as an email to your company cell phone. Without opening the email, one could discern from the subject line whether an immediate response was required, and if not, ignore the occurrence and allow the SMS process to deal with it in due course. People had become so attached to this system that they associated the time of message delivery with the time of occurrence, not realising that sometimes the message regarded an event occurred days ago (e.g., delivery failure or the duty manager was so busy at the time that he/she had "risk assessed" the message and put it on the long finger).

Such was the outrage that I had presumed to remove this crutch, upon which so many people actually based their safety management practice that I was forced to suspend the project and go back to the drawing board. I had to ask (and pay) the developer of the software to build a new piece of code to generate the identical SITA-type message and email it automatically upon report submission. I did manage to salvage the benefit of being able to control the distribution list, and the duty managers loved me, but the practice of managing safety by reference to these emails was perpetuated.

I was forced to concede defeat and revert to the trenches. As I licked my wounds, I came to realise that I had utterly failed to convince only a small cadre of intelligent, motivated managers of the benefits I thought were blindingly obvious. What I then understood was that, in fact, I had failed to understand the culture, where safety was almost exclusively reactive: if the signal indicated a crisis or need for urgent action, then everyone was on the same page. Otherwise, it was ignored, and at the end of the month when it came to reviewing safety performance, these same people were looking at all the safety occurrences for the month for the first time. The only ones they remembered were the ones that had caused them to open the email. Everything else was left until later. We needed to change and become more dynamic and proactive.

I also saw clearly that the presence of the email signal and the fact that I now was forced to publicise its presence as a documented part of the process (we now live in a world of documented, implemented and transparent!) was a blatant deterrent to reporting. Pilots and other staff started coming to me directly, asking to submit reports that would bypass the email stage, to avoid telling "the world and his mother" about the error they had made, or the situation they had found themselves in with their crew. Nobody objected to submitting bird strike reports, sure isn't it always the birds' fault, but self-reporting of errors and hazards was under threat.

I completely revised my change management strategy and explored all the possible benefits of a system that would provide managers with timely, relevant, prioritised safety information. It would be presented in a format that respected managers' hectic workload, but also in a way that facilitated their staff in managing the information efficiently and effectively. I sold the idea of uninhibited reporting, the benefits of facilitating staff reporting directly. I reminded them of the reduction in workload on the Operations Centre, allowing them to concentrate on keeping the business running. And I convinced them one-by-one that the change was actually a good thing, that it would help them and make life easier for them.

The changes were made to the system and rolled out without fanfare. New daily reports were designed, easily read and delivered every morning at the same time. Staff were told how the system was now designed to protect their identity, to make it easier to report in confidence, and the results were dramatic. We saw significant increases in voluntary reporting, so much so that managing the volume of data became a threat, but a welcome one as our insights into operations increased significantly. We educated staff about the difference between communicating and reporting. If you are standing in the grass looking back at your aircraft, it's probably not the time to be looking for an iPad to submit a safety report. Pick up the phone and tell the company. But if you struck a bird, suffered no damage and after proper inspection, you are on your way again, well, concentrate on flying the aircraft and fill out the report later or at a quiet time.

The funny thing is that to this day, many people still refer to raising an Incident Signal when they submit a safety report. I thought I could change direction because it was easy. I did not anticipate leaving important people behind. As the ocean liner, turning is possible, but it will cause damage and upset. Planning and anticipation are critical to a smooth and successful change.

My messages to the reader: clear identification of the direct benefits to stakeholders is key to successfully implementing change, and the same applies no matter the size of that change, from small adjustments to SOPs to full-scale culture shift. People like to know what it means to them. I won't forget this lesson.

Note

1. U.S. Department of Defense (2002). News Transcript. Retrieved from https://archive.defense.gov/Transcripts/Transcript.aspx?TranscriptID=2636

Chapter 6

Is Safety Part of Your Business Model? Turning a 'Simple-to-Fix' Safety Incident into an Opportunity for Everyone

Derek Stevenson

Contents

My story begins with another person, the Health and Safety (H&S) auditor, whose day was no different from any other. His objective was to undertake a safety audit while an engineer was working on a rooftop and undertaking telecommunication engineering duties for the 4G rollout of an Operators Network. The project delivery was aggressive, with work being delivered to between 40 and 60 sites per day under a multimillion pounds contract between the principal contractor and the client, a telecom operator.

The H&S auditor witnessed an engineer climbing a 4.8 m fixed ladder to access the roof where the work was being performed. The engineer was unattached from the fixed ladder while wearing a safety harness. The engineer was in breach of the Working at Height Regulation 2005 of the United Kingdom (UK), together with the collective working at height policies of the client, principal contractor and contractor for whom the engineer was working. I, as the H&S manager working for the principal contractor, was alerted to the incident by the H&S auditor.

The salient point from the day was that the job was stopped, and the engineer requested to return to the contractor's offices where he discussed the incident with the project manager. The agreed outcome between them was to provide the engineer with new climbing Personal Protective Equipment (PPE) with an accredited inspection certification and move onto another project the following day; the PPE issues pertinent to the incident will be described. The interesting point here is that for the project manager and the engineer this seemed a suitable and sensible solution to the morning's safety problem: issue detected, issue fixed. Unbeknown to them, I was present at the said offices and witnessed the interactions between the engineer and management over the incident. This unplanned encounter influenced my subsequent investigative thoughts. To me, the investigating officer and the H&S manager responsible for maintaining and improving the safety standards of the project, this oversimplification of the incident's causality was galling.

Nobody closely involved in this incident had learned anything from the conditions it presented; a similar incident could occur without such a favourable outcome. The approach and action of the contractor's project manager only served to reinforce the engineer's approach to misbalance safety and productivity, as elaborated below, by setting the latter as the highest priority and possibly adversely affecting his livelihood. The organisational policies and values contained within the contractor's business management systems remained dead words and had been undermined. Instead of understanding the causes of this high-potential near miss, the contractor's senior management team was presented with an internal statement articulating that the incident occurred due to the engineer's careless and the problem was resolved swiftly by retraining the engineer and allocating him to another project.

When I started collecting the relevant documentation, PPE inspection certificates, Risk Assessment Method Statements (RAMS), photos and interviewed the engineer involved, further failures became apparent. The typical accident investigation process I used to follow until that time would not be suitable for this case. The failures, if considered in isolation, would have normally triggered a binary-type investigative journey, which would result in a prescriptive outcome to rectify the fault or issue as exactly happened between the engineer and the project manager. I decided to turn to what I learned from my postgraduate H&S studies. By selecting to apply a Fault Tree type investigation to the incident, active failures could be mapped as combinations of deeper failures. I chose to apply this type of analysis because I had practised it during my studies and, thus, I felt comfortable with it. Its application to this incident revealed a far deeper and systemic cultural failing than the easy-to-spot finding of a negligent engineer forgetting to put his lanyards on his harness and attaching them to the fixed ladder. Below, I explain what I found along with my understanding as an insider of this specific industry sector for many years.

When the H&S auditor witnessed the engineer climbing the ladder, the engineer was wearing a full body climbing harness with an attachment for a fall arrest climbing lanyard, but the latter was missing. To me, this was an important point; why would an engineer just wear a harness? It serves no purpose other than being cosmetic if not attached to a robust, structural and solid point. This question was answered when I noticed a single lanyard laying in the engineer's vehicle because the other one was away for repairs. Thinking reasonably, the engineer could not remain physically attached while climbing the fixed ladder by using one lanyard alone. The engineer must have dealt with a dilemma, I extrapolated. This particular point is supported by the fact that the engineer, who is a subcontractor to the contractor, was supposed to wear climbing PPE because the height of the ladder was greater than 3 m in vertical length and any fall would cause injury. Having only one lanyard always compromised his safety as he remained unattached when working at height.

My understanding was that the engineer's judgement was influenced and compounded by the possible implications when work could not be completed. If any job was not aborted in agreement with the client, the costs could not be invoiced through the supply chain (i.e., subcontractor/ worker – contractor – principal contractor – client). Hence, the engineer could not claim payment for his days' work if the client and/or the principal contractor had not consented to rejected work. This was further intensified by a 90-day end-of-the-month payment term by the client, which cascaded through the whole supply chain. In essence, the client was driving the whole supply chain to live on borrowed credits, which could lead to cash liquidity issues and commercial decisions influenced principally by cost and, possibly, compromising safety. This model is typical in construction activities, where the commercial risk is pushed down through the supply chain to manage costs and ensure a budget without or only limited overruns. This, of course, does not mean that safety is always at stake; it is rather about the mechanisms each company employs to deal with the trade-offs imposed by any prevailing business model.

The engineer, by having only one lanyard, was effectively putting his live-lihood at risk until this issue would be resolved. I believe that he decided to wear a harness, even unattached, as it was highly important to him to work and earn his living. Nevertheless, he might have convinced himself that the risk of falling from height was small as he had undertaken similar tasks plenty of times before without an incident. Furthermore, as my investigation revealed, the engineer's training for accessing rooftops did not include any component for accessing heights through a vertical ladder of this length. The question, of course, remains unanswered: had the engineer undertaken such rooftop access training, would he be able to identify the significant risks posed and flag the issue to the contractor's project manager? As you will understand from the rest of the story, I personally doubt about it.

Moreover, the engineer was not backed and empowered by any contractor's policy to make a safety stand. Thus, the engineer could not challenge the commercial status quo of the contractor. He was bearing all the risk and knew no different than the knowledge gained from inadequate training together with an incomplete RAMS that missed the significant risks related to the hazard of falling. My investigation additionally revealed the following:

- The engineer had not been given a replacement PPE.
- The engineer's harness and lanyards had not been inspected for ten months, whereas the UK law requires an inspection every six months.
- The Safe System of Work had not identified at the planning stage any requirement for a rescue system, on-site monitoring and two-person attendance when accessing a rooftop site through a fixed ladder.
- The RAMS had not identified the point above either and its provisions were very generic.

I reasoned that these behaviours probably ran throughout the whole contractor's business as an 'action slip', as defined by Reason and Hobbs,[1] where the action runs on rails along a familiar route – the routine of the task – but this time because of the change in circumstances an 'action slip' occurred. Had the information of the two-person site been available during the planning process together with respective training, assets and inspections, both the project team and engineer could have made an informed decision.

During subsequent meetings, when the contractor's senior management team confronted with the depth and breadth of the findings discussed above, they acknowledged that fundamental failings existed within their business. They agreed on a programme of change including (1) a review of the core training to ensure it was fit for purpose and matching the job profiles; (2) consistent PPE inspection and logs; (3) addition of site-specific details to RAMS; and (4) timely project planning and delivery of a workshop for all operational staff to become familiar with the values, objectives and processes of the business. Below, I provide more details about the specific changes.

First, after collecting data about the most serious breaches, I involved the workers in devising agreeable ways to manage hazards instead of me using a 'big stick approach'. In practice, this was an example of a transformational approach accompanied by the sharing of data with the H&S leaders of the various contractors. Also, I requested that all engineers had the required skills to undertake any tasks defined as critical and I compiled a list of qualified persons that was checked daily. The records on my list also included the availability of climbing PPE, rescue equipment and calibrated test equipment. This way, I could check whether everyone on-site had completed their training and had suitable and available means to carry out their tasks. Moreover, as part of a 'go list', I knew that the critical parts of the project, including the weather conditions, had been accounted for on each worksite; thus, I was able to limit aborts and critical safety failures.

Furthermore, I changed the RAMS from generic to site-specific by including emergency procedures, Controls of Substances Hazardous to Health (COSSH) and environmental considerations such as waste and spillage management. Briefly, each RAMS aligned with construction activities and drew on site-specific, task, construction and dynamic risks to make it a live document that the engineers used and referenced. Last but not least, I linked Contractors' reported accidents and incidents with their site inspections, audits and leadership visits to identify common themes, trends and emerging threats. This was communicated weekly to the contractors using a conference call. This strategy was the process driver that brought buy-in from the contractors' H&S teams and rippled its effects across the project, which resulted in the reduction in the Accident Index (AI) score of the client.

To ensure the substantial and successful implementation of the above, I performed audits and unannounced inspections. Now, I am in the position to say that by implementing the 'talk of policies, procedures and processes' throughout their business so that the organisation 'lived them in everything they did', the contractor became operationally safety-balanced. This is a credit to them and the industry. The H&S performance improved by taking a collaborative approach across the whole project and sharing the same H&S goals and challenges.

The first lesson learned from this near miss is that the H&S practitioners must be able to change their 'standard' safety investigation methodology and trajectory as they uncover evidence and facts. Something might initially seem straightforward to conclude; however, a different investigative methodology may pull the layers away, reveal complex underlying issues that harbour the true cause(s) and produce completely different outcomes. Often, by leaving aside for a while your typical daily activities, becoming fully immersed into the safety investigation undertaken and trying different methodological approaches leads to testing your hypothesis from different angles. When I applied my knowledge and used investigative skills different from my usual practice, I was taken on a different journey.

Second, any near miss, instead of being tagged solely as a lag indicator, can become a useful leading indicator if you focus on the opportunities to improve your management system by finding and rectifying the 'gaps in the defences'. The findings from one investigation could be applied elsewhere across the business where you see similarities and a similar story in the data. In closing this story, I want to state a last reflective point: obtain confidence in understanding your own thinking, become able to articulate that thinking to others and produce facts and evidence to support it based on research and data. These steps will render your changes credible and valid and will signify positive safety leadership.

When Profits and Rush Could Override Safety

It was a mid-afternoon in July. A principal telecom contractor had engaged a lift subcontractor to help deploy a 30 m temporary telecom tower for a sporting event. During lifting operations, ballast plates, weighing one and a half tonne each, are used to stabilise the structure. They can become unsteady when an external force is applied, a condition that induces swaying and introduces oscillations into the lifting accessories and jib arm. That day, when lifting a ballast plate, the load started swaying violently from side to side. The workers 'hovered' the load to gain control, but it struck a landed ballast plate. The lifted ballast plate fractured, the lifting accessory known as 'stelcon hook' slipped and the plate became unattached, fell and broke into five large pieces.

An electrical subcontractor, who was employed to connect the generator and provide the earthing once the tower was erected, was working below this lift. During the event, he was hidden from view because he had crouched down among the landed ballast plates. He got struck from the plate pieces and suffered serious injuries requiring an air ambulance to evacuate him to the hospital where he remained for several weeks. The police and the Health and Safety Executive (HSE) were informed and attended the scene, and a criminal investigation proceeded under RIDDOR.[2] Once the police were satisfied that no criminal intent was manifested, the investigation was handed over to the HSE.

The principle contractor instructed an H&S consultant to undertake the initial accident investigation as, at that time, they did not have a full-time employee to manage their H&S affairs. I was employed later as the H&S manager and undertook the accident investigation proceedings. The plan was to review the collected documentation and apply the Bow Tie[3] and Five Why[4] methodologies on the event's timeline. Based on my previous investigation experience, I was aware that peeling back the layers often reveals themes that are not readily apparent during the initial investigation. What I became additionally aware of during the specific investigation were three major contributing factors, which I explain hereafter.

First, the design of the ballast plate was a concern. The principal contractor had been underwritten by a parent company that supplied the ballast plates from Europe; the plates were preconstructed and shipped to the UK. The European manufacturer was producing plates of this type for the last 16 years. Interestingly, the British Standard, BS EN 10080:2005 'Steel for the reinforcement of concrete, weldable, ribbed reinforcing steel' gives no actual specification or figures, which are left to the National Standards. Therefore, the Standard NEN 6006:2008 'Steel for the reinforcement or concrete' applied to the ballast plate and the accompanying stelcon lifting chain accessory. The supporting documentation stated that the ballast plates' construction conformed to the standards by using steel rods in precast concrete. However, the investigation uncovered that the fractured ballast plate had no rebar framework included in its construction. Therefore, the ballast plate did not conform to the NEN standard above, although the certification provided suggested otherwise. If the rebar cage had been fitted, the ballast plate would largely remain almost intact during impact.

The attachment of the stelcon lift chain to the ballast plate was achieved by using friction to exert the required pressure on the securing slots contained within the ballast plate to lock the lifting chain in place. The ballast plate could be secure if an equilibrium triangle had been achieved through correct angles between the lift chain and jib hook, correct lifting force had been exerted on the stelcon hook and slots and the correct length of lifting chain had been used. If an outside influence exerted another force upon any of these three parameters, then failure would occur. Often, after rain, the locating slots in the ballast plate would be full of water that would be displaced when the stelcon lifting chain was inserted. If the equilibrium triangle was not balanced, because of the poor coefficient of friction between the stelcon lifting chain and ballast plate slot, then slippage would occur. The manufacturing company had chosen this design because it allowed the ballast plate to lie flat without any protrusions or lever points to fracture or topple the ballast plates when they were stacked upon each other.

Consequently, the lack of a rebar cage, whose presence had not been ensured by the principal contractor, and reliance on friction and applied force to secure the lifting accessory was a latent hazard because the load was not physically secured. Retrospectively, it seemed reasonably foreseeable that the load could fall if the equilibrium triangle was not maintained. Prospectively, only long and rich operational experience accompanied by site inspections can lead to the detection of such flaws.

The second major factor relates to issues with site management as it will be understood after I explain some standards practices in this sector. Given that all subcontractors are paid either by the day or per piece of work delivered when the overall operation is fragmented, and the subcontractor is needed only for part of the day, the cost for the whole day is still invoiced. Briefly, the subcontractor is booked for the whole working day to undertake the assigned job regardless of the actual time and duration of involvement, which within UK telecoms operations is considered as standard practice. Thus, according to the typical contractual clauses, there is no opportunity for any subcontractor to go elsewhere and turn up later at the time of requirement. In the case of breaching this clause and the subcontractor fails to attend the worksite or its absence causes delays, the subcontractor could be liable for all the costs of that day.

Although the clause described above is in place to ensure all resources (e.g., staff, materials and equipment) are on-site when required and project costs are within budget, it pushes collateral risk down the supply chain. While effective in project delivery, the problem is that subcontractors rush to complete their works by occasionally comprising the safety of others (e.g., staff working at ground level installing and commissioning telecommunication equipment while riggers are working on aerial platforms to install antennas). Everyone wants to finish the assigned job as soon as possible so that they do not delay the overall project and upon delivery can move to the next one.

Such trade-offs become even more unbalanced when the on-site managers of subcontractors have strong personalities and exhibit selfish behaviours to accomplish their tasks and leave the worksite at the soonest possible time regardless of the degree to which safety is compromised and negative side effects are imposed on others. This relates to the widely spread perception that 'smashing the work out' is a positive behaviour. Even worse, some contractors form small groups with strong bonds and bully others who they see as responsible for delaying the completion of their works. Consequently, it is imperative to provide strong safety leaders at the site who are empowered to manage the site as they see fit, especially during complex works, to assist, offer guidance and challenge behaviours.

The situation becomes even more complicated when considering that the Person in Charge of Work (PICW), who can theoretically demonstrate overall leadership and exercise coordination, works at the same time as a member of a specific team. Modern business models dictate creative accounting to invoice for line items when they are not used or achieve a small profit to compensate for the competitive tender price to win the contract. Thus, a contractor often offers a price for two roles at the cost of one. An investment in a separate PICW role free from requirements to participate in another job would discharge the H&S duties of the contractor. I believe the outcome of the specific event would have been different if a paid and skilled PICW had challenged the electrical subcontractor who was undertaking works while lifting operations were being performed.

Another interesting point is that, according to the event timeline drawn during the investigation, there were still many working hours left for the electrical subcontractor to complete the tasks during daylight and travel home. I reasoned that the visual cues of the lifting operation and subsequent mental calculations of the electrician added up and developed his belief that he had arrived on the worksite later than planned, and, hence, he had to complete his contractual commitments the soonest possible. This internal conflict must have dominated the electrical contractor's mind who committed to his tasks without even seeking approval or checking the proximity of other operatives on-site.

Given that the electrical contractor had become rather despondent, could he have requested the lifting operations to stop temporarily so he could complete his task safely? In this case, the PICW, ideally, could have produced a revised work plan to allow the electrical subcontractor to undertake the works. However, should the electrical works overrun, the principal contractor would be liable for the extra abortive costs if the lift subcontractor could not complete their works during that day. The abortive cost for a lift contractor was £850 over the £350 for the electrical contractor. Therefore, it is reasonable to assume that the PICW would prefer to give priority to lifting operations instead of delaying them. Maybe, this discouraged the electrical subcontractor from requesting a temporary pause of the lifting operations.

Poor documentation was the third major factor my investigation revealed. The RAMS were generic with little detail about site-specific and dynamic risks or onsite management. As task complexities increase and risk multiplies exponentially over time, detailed risk management is required. In this case, appreciation of the complex risk was missed. A project manager wrote the document and the managing director had underwritten it, both with limited H&S training. Indeed, no safety professional was employed to provide advice.

While the pathway for improvement to learn and prevent a reoccurrence was available during my investigation, I felt that my thinking was shackled with the on-going investigation being performed by the HSE. I was not 'free' to peel the accident layers away with the threat of being accused by my co-workers and the managing director of being coercive. I was not able to share my findings, being open and honest while the formal HSE investigation was proceeding. This put me in a position of defence to limit criminal liability and its effect on the company and its employees. Only once the HSE investigation was completed with the final conclusion published, one can resort to function as a safety investigator to learn and improve the Safe System of Work. This situation does not mean that the HSE investigation would be unfair by default or would not uncover the same or more flaws in the system. Nonetheless, it is common across the industry to worry that someone external cannot fully understand the context of work and can focus on the persons and evidence proximal to the event, and, unintentionally, miss the contribution of other critical factors. Maybe, a joint investigation with the HSE and the company would be a solution to this perceived threat, but this is not the practice to date.

Therefore, in contrast to the first story where I had the opportunity to improve the system, and the aftermaths and messages were easier to communicate and resonate with the business, during the investigation of this accident, the shadow of possible severe liability hindered the process from improving and rendered everyone defensive of their decisions and deeds. An accident investigator must be mindful that each investigation is different, and even with the best of intentions and skills, unobstructed execution of investigative tasks cannot be guaranteed even if any perceived threat to the investigation process is real or not.

Notes

1. Reason, J. & Hobbs, A. (2003). *Managing Maintenance Error: A Practical Guide*. Farnham: Ashgate.
2. The Reporting of Injuries, Diseases and Dangerous Occurrences Regulations 2013. UK Government. Legal requirement to report certain injuries to the enforcing authority – the Health and Safety Executive.
3. Khakzad, N., Khan, F. & Amyotte, P. (2012). Dynamic risk analysis using bow-tie approach. *Reliability Engineering and System Safety*, 104, 36–44. DOI: 10.1016/j.ress.2012.04.003
4. Wikipedia (n.d.). Five whys. Retrieved from https://en.wikipedia.org/wiki/Five_whys

Chapter 7

The Development of Mental Health Proxy Teams and a Relationship That Threatened the Quality of a Safety Investigation

Dimitrios Chionis

Contents

Almost ten years ago, I joined an airport's staff as a psychologist with a focus on mental health. There was always so much to do and manage within the airport including field observations, monitoring, and safety briefings. The issue which was torturing my mind was that all these activities around safety felt like falling in the void, like no one was actually caring, or even listening to me. At first, I thought I was inefficient due to my lack of experience, that I was not capable enough to communicate the right message, to the right people, in the proper manner. My stories relate to my roles both as a psychologist and as a safety investigator.

Why a Little Motivation Can Lead to a Big Success

Due to several safety incidents during that period, the aviation authorities became more focused on mental health issues, mostly regarding the recruitment process and mental health check-ups which, nonetheless, were part of the annual employee check-ups. My duties were to monitor people who had previously been under treatment and were fit-for-work, consult newly hired personnel concerning adaptation issues, and assist with lectures on safety issues with an emphasis on human factors. The problem at my hand was that I felt that I couldn't effectively manage all the personnel in the airfield with a workforce population close to a three-digit number. So, I laid down my plan to multiply mental health monitoring and awareness in the airport by creating mental health proxy teams.

Specifically, I aimed at combating the unawareness of mental health issues among peers and between front-liners and managers and support the idea that all people should have access to all the possible help they could get, especially for mental health. Also, I believed that I could manage it until the end with excellent results. A timeline of three months was set up with one goal to be achieved per month to consider the endeavour successful; the three goals were Knowledge Foundation Building, De-stigmatisation of Mental Health as a Field and Communication Skills Development. These three goals constituted a tailor-made programme as I saw it fit for the airport and the people working there. Below, I explain these goals more specifically along with the context within I had to work, and I elaborate on the strategies I employed to achieve the specific goals.

1. Knowledge Foundation Building

   Although I had been at the airport for five years, I noticed that people were seeing me as a foreign object in a hard-industrial environment. Due to organisational stagnation of standardising my job description for some time, I had to take a lot of initiatives to introduce myself and the profession I represented in an environment lacking one so far. I cannot attribute that only to the organisation, but also on my hesitation as a newcomer. A major barrier which worsened things was the hardship of stigma coming from other health professionals in the airport, medical doctors to be more precise. A more welcoming approach was from those who thought that psychology was more like popular science or even a strand of the paranormal. Most of the airport staff were laymen, blue-collar workers with an average age over 40 and did not worry much about the "academic stuff", as they used to say.

My first objective was to inform employees about the whereabouts and the job description of mental health professionals. Equally important, my second objective was to offer useful information about proactive mental care and effective communication. The third objective was to explain the various approaches of psychology towards human behaviour, the reasons for behavioural deviations, and possible explanations about the issues staff had experienced in the field. For example, some colleagues had shown symptoms of depression. I had to clarify among each other the specifics of depression, its symptoms, and possible manifestation of these symptoms during work.

2. De-Stigmatisation of Mental Health as a Field

Regardless of building knowledge, the fear of a stigma of those who had asked help or the fear of being "one of them" if someone had considered asking for help, were still present and powerful. And how do you combat fear? One approach is to show that fear is no more than **f**alse **e**vidence **a**ppearing **r**eal. In mental healthcare, stigmatisation might reach the caretakers as well, especially in environments rich of stereotypes about everything that is not technical, mechanical, or, generally, tangible. The hardships during the pursuit of this goal involved higher pressure because as a psychologist I was not viewed as a member of the health team in the airport. A frequent comment from colleagues I was trying to approach was "How come you want to talk about battling stigma when you can't deal with the medical doctors?" Indeed, I had to lead by example. Due to this issue, my original three-month schedule for the delivery of one goal per month changed to six months to show to staff that I could do something about the situations I was facing. The knowledge foundation alone as per the previous objective was not enough to convince them that their time was worth spending with me, and they were right.

3. Communication Efficiency

The last touch was the building of communication skills. Some may have a natural talent in storytelling, and some may need to struggle to read out in public even one line of a poem. To achieve this goal, I delivered some workshops about body language recognition. These workshops indulged the attendants to observe and interpret others as well as engage effectively in conversations; to be able to recognise and interpret non-verbal and body language of others, they had also to know theirs. Also, during my discussions with employees, I noticed stutter or use of conversation fillers, such as "um" and long, unintended pauses. To counter these phenomena, we tried to practice small talks and casual conversations and demonstrate how one can oscillate around an argument or empathise with others. The form of all communication practice was in a storytelling[1] mode since the aim was to stimulate critical thinking over the communication of a field that the participants had just gained insight.

The inspiration of how to lay down my approach to achieving the three goals described above came from Heath and Heath[2] who stated that "to change behaviour, you've got to direct the Rider, motivate the Elephant, and shape the Path. If you can do all three at once, dramatic change can happen even if you don't have lots of power or resources behind you". I chose this approach over social theories (e.g., Terror Management Theory) or other behaviour-based approaches (e.g., conditioning) because it included a unilateral approach taking into account equally the environmental factors around the person you are trying to assist, as well as personality, emotional and cognitive traits. Concerning the tactical steps, let us take one candidate from within this project as an example to explain it further and use the name "Jim":

The first step was all about "shaping the path" by building the habits and modifying the environment. Jim was working for a technical shop, and he was very creative, quite enigmatic, friendly, spontaneous, and practical. In the workplace, Jim's habits could colour him unpredictable and sometimes frustrating towards others due to difficulties in having a normal conversation (i.e., easily agitated from minor issues to get a kind of revenge on someone who was harsh on him the day before), reading the other party's body language and empathising without getting bored. The same frustration hit me many times with him saying "Come on doc, people don't care about these things and they shouldn't, we just build things here with our hands, and that is enough". Based on his good sense of humour, we first worked on the diffusing-the-tension part of communication. From the same root, we cultivated the practical recognition of stigma towards mental health and empathy through locating the reasons why the other party in the conversation could be inclined to stigma. As such, we built the habits and the appropriate communication environment in the technical shop.

The second step was about "motivating the elephant", finding the roots of fear and instil growth. Jim's inhibition about all things he couldn't grasp and fix with his hands was astounding sometimes. Following a few discussions with him, we discovered that he lacked focus towards what mattered in each conversation. Jim wanted to have a quick-fix solution to all problems during conversations, which was exactly the opposite to his practice when he was fixing a machine and wanted to invest time and attention to carry out his tasks safely and with high quality. Thus, his difficulty in detecting the most important parts of discussions with other colleagues was rendering any verbal interaction in the workplace pointless to him. So, we fixated on locating the key messages in all conversations and how he could relate to them, and, if he was unsure, how he could raise questions to achieve a more in-depth understanding. Surprisingly, it was more fun to him when he managed his way around that point. He was able to handle minor insults or hard jokes without holding a grudge, he stopped stuttering when he was getting nervous, and he was able to humour back eloquently (i.e., "Hey Jim, you should work more on those hands on the mock-up bench man, you might be missing a finger already". – *"No thanks man, I can always borrow one of your fingers, just let me get my chainsaw"*).

The third step was about "directing the rider", scripting the critical moves, and pointing the destination of how to communicate with others. This step had to connect with all the goals of this endeavour. Jim had to be able to use his knowledge, disclose the reasons behind the avoidance of stigma of mental health issues in the workplace, and help others do it as well as a proxy communicator. The difficulty in this step was the existence of some colleagues with depression within the technical shop. So, Jim had almost the same level of hardship with his colleagues that I initially had with him. The critical moves here were to use the conditions towards our advantage and elaborate on the positives during the working hours. For example, depression kept two colleagues away from work, but they were back on their feet after three months and efficient as before at their posts. The satisfying part was that Jim, who was never close to those two before, managed to change the climate and bring along one more colleague, and they started hanging out after work as well.

Of the initial 50 people who found the idea "interesting", only 14 followed all three steps to the end, and Jim was one of them. To my disappointment, retrospectively speaking, I was too stubborn to see that I was going too fast and too harsh on topics people were not ready to follow. For instance, there was the case of "George", who was too reluctant and arrogant towards the whole program, thinking that it was something he could do by using some self-help, popular science, psychology handbooks, and so on. Because he was constantly comparing what we did with the contents of his books, he decided to drop out and not bother himself with "these things" anymore as they seemed "too soft" for him. What I would do differently is to run a sentiments awareness crash-course. Personnel from the airport saw sentiments as a weakness, as something strictly personal, not being able or just refusing to explain or describe. Ajzen's Theory of Planned Behaviour[3] (TPB) could be an approach to fight against this, and I believe that with a little more dosage of improvisation and less eagerness to be exact on the steps and time-schedule from my side, it could have worked. TPB is a stricter behavioural approach which can be applied to modify more concentrated patterns of behaviour. According to my understanding, feelings fall into this category, as TPB aims to modify the behaviour or its lack thereof.

Beyond this mildly intrusive course of action, a moderate coaching approach could have also been applied. Taking the case of George, for example, an indirect and self-management approach could have worked better, especially when people project their negativity in the form of arrogance. A coaching self-management approach would include a general approach partitioned into life coaching (i.e., to counter unsettling life events, such as a nasty relationship), career coaching (i.e., to counter deviant decision-making on the chosen career path), goal-setting and time management (i.e., to counter pointless workload increase), and self-discovery (i.e., to aid inward insight of themselves and aid self-appraisal and self-recognition). The course of action I followed involved increasing resilience about mental health stigma and build team-working skills by increasing motivation to connect with other co-workers, as in the case of Jim. The issue with that approach was that it was too lenient, and it resided into too simple tools, such as the typical Strengths – Weaknesses – Opportunities – Threats (SWOT) analysis. The feedback received during the application of the intended program revealed that complexity in the thought skills boosted interest and self-appreciation since participants felt personal progress. The startling fact was that the less they were thinking of themselves, the more they were eager to acquire new communication skills and techniques. My interpretation of this is that over-compensation is stronger among those with low self-esteem regardless of their actual skill set.

The bottom line is that safety practitioners, not only psychologists, should always consider and grasp the needs and the "buttons to push" of the people they are dealing with regardless of their role as top managers or first-liners. The important point is to adjust before applying any approach. You have to adjust your behaviour towards the mini climate of the organisation you are working and for each one you wish to approach. Moreover, this is not a skill built in a day; it takes time and valuable energy. There might be some barriers, either people, procedures or stereotypes, which will make the job even harder. My advice is to manage any possible hostilities or aggressiveness rather than react to them; build alliances and separate the people from the problems, and through this you can see how a little motivation can lead to a success story.

How Could a Strong Friendship Make You Blind to Facts?

It was evident that during my service at the airport, strong ties among staff with other specialities would arise. Human relationships often provide various proportions of joy, drama, and frustration, and they can be highly rewarding for personal development. The implications begin when the official status of someone inside an organisation interferes with personal affiliations within the same organisation. As a member of an accident investigation team, it was a hard moment to be assigned on a case where one of the persons I appreciated the most was the major actor. "John" had a high managerial position at the airport by the time I first got there, and he was among the first who supported my work as a psychologist. I admired him because of his modesty, good manners towards everyone, and his eagerness and readiness to listen to all sides during an argument or any formal proceedings (e.g., audits, review meetings).

John was involved in an aircraft accident, where he was the pilot. When I first heard it on the radio, I froze because I thought "John is gone". However, the situation was a bit more complicated, and, gratefully, John was still alive. That Wednesday morning, John was scheduled for a certification renewal flight, and he was flying with another highly experienced pilot in the cockpit. The thrill of the moment caught John, and he wanted to take some cool photos of the sunset near the mountains south of the airport, close to the sea. And there was where the problems began. The photo-making was a last-minute decision during the flight. The crew did not have enough time to plan a new course, put new stirring points, and check all maps in detail. It was common knowledge that those mountains have high voltage cables at a high altitude. John wanted to take a cool shot of the sunset while "knifing" the aircraft between two mountaintops. So, as he was passing through, he chipped off one-third of his left wing due to contact with a high voltage cable in the area. That aircraft part was shot out at the nearby beach and ended in the sea. He managed to level off and do an emergency landing at a nearby small airport. Any substantiated doubts of the crew about the existence of the cables and the exact sequence of decisions and actions before and after the accident remain yet blurred. In my mind, I was vouching for the middle truth; John began the knifing to avoid the cables at the last moment, while he was flying low, almost at the lowest permittable limit (300ft). This underlines how positively I was predisposed towards John.

Without any delay, the authorities formed an investigation committee to unveil the causes of the accident. The interesting part was that I was involved in this case because I had just graduated from the investigators' course, so I had the most updated knowledge. Also, the crew had survived the accident; thus, the idea was that a human factors specialist should be included in the investigation committee. I already had a terrible initial feeling about the procedure which started in a totally different manner than what I had been taught during my recent course. On the other hand, I was well aware that such training usually deals with 'ideal' conditions most of the time (e.g., everyone is knowledgeable of her/his role and responsibilities, everyone is following the protocols and investigation procedures). In other safety events, a human factors specialist had not been assigned as a member of the investigation team, even if it was required by the respective investigation regulation. It was unclear why in some cases a pilot, an engineer, or another specialist could undertake the investigation of human factors without formal training, while in the particular case, I was appointed as such. I felt that I was needed as a tick in the box to show that there was a human factors specialist in the team and the investigation results would be credible.

Next, the investigation began. We had to take about 20 interviews from various specialities working that day at the airport and analyse this information. The analysis part, however, fell only on me for some "you are the rookie" reason. That behaviour almost made sense to me in a way I must admit. At least I complied to adjust better among the other investigation team members since this was my first investigation, and I wished to be appropriate and present a good image. Our team used the Human Factors and Classification System (HFACS) model during the investigation, and we excluded any intentional handling to damage the aircraft or anyone and anything else. The HFACS model was part of the material taught in the investigators' course and widely used by the organisation. Whatsoever, the nature of the accident at first glance looked like a straight violation of all rules of safety; John just went and did what he did, and the co-pilot never stopped him, neither even raised concerns.

Looking at the findings, the human error, in this case, was not that easily located as it seemed in the first place. There were contradicting opinions in the investigation committee concerning the issue if it was an exceptional violation or not; if it were, it would not be our responsibility anymore, and the human resources and/or security department should take over. Additionally, another debated issue was the degree of organisational influence as a preconditioning factor and, if so, the corresponding personnel involved. I persisted that for the kind of flight we were investigating John had simply made a decision error and not a violation. This position was probably influenced by my affiliation with John, but it was also supported by another team member who did not have a similarly close relationship with John. After analysing John's moves retrospectively, the other investigator supported that John had followed the particular sequence only to avoid obstacles in his course due to the last-minute change of the latter without enough time to account for the terrain and the wires.

One of the hardest parts was that even if I wished to push the investigation to a direction that it wasn't John's fault and the accident was a strong example of organisational flaws, after connecting the dots, I realised that John was both under organisational influences and personal overcompensation for past events; it was a case of him attempting a last glimpse of his "strong flying" skills as a pilot. I could almost sympathise with John because he used to be the best pilot at that specific airport more than 15 years ago. Circumstances and alliances within the pilots' group managed to send him away after he was charged with a similar accident back then, although it was uncertain if this past event had happened as it was documented due to a lot of contradictions in the findings. However, I could completely understand the insistence of the investigation team to get back to that case. I would probably have done the same.

As the investigation was progressing, I felt more and more unpleasant and dissonant with myself. I was too biased towards John, who was the major actor in the incident. The awkward moments came one after the other; first when I was forced by the chief investigator to be present during John's interview. I tried to find an excuse, but the chief investigator insisted very convincingly by saying "so is this how you want your first investigation to be?" Later the same day, John met me by chance at a local bar, and I was unable to look at him straight to the eyes and hold a proper conversation because

I was feeling so embarrassed. John had supported me so many times, and now I was part of a committee which could mean the end of his flying career. If I could have done something different back then, I would have asked from the beginning to be excluded from this situation because of personal affiliations. I must admit that I was afraid to deny my appointment as an investigator and I also felt excited about my first investigation. But I was imagining something different than investigating friends. During our training, we were warned that this course might lead us to a morbid situation, but my excitement overrode the knowledge acquired during my training.

During all phases of the investigation, I was too hesitant to reflect upon the facts, and I could have contributed better concerning the behavioural modelling of the involved pilots and the contribution of the organisational culture. When I studied the files from that old accident and compiled the flying profile of John according to his records, subject to hindsight bias, it seemed like he was going to do something dangerous one day or another. Yet, I also had the urge to show excellent work to the chief investigator with the least consequences on John's future. However, a fellow engineer helped me with that by advocating that John's flying capabilities kept the aircraft flying and he managed to land. We modelled that John was not aware of the wires, but he did manage to notice them and bank to the right, just enough to have only one-third of his left wing cut off. Therefore, we had underestimated the ability of John to adapt even within the course of an unfolding failure. The problem is that if I did the investigation again now, I would say that John had seen the wires, he wanted to avoid them, and continued knifing through the mountaintops; it is still hard to overcome my favouritism to John as a friend and colleague.

At the end of the investigation, the chief investigator was content with the results, as was the higher authority. However, for me, I failed that investigation. On the one side, I was a rookie investigator trying to show a good image to earn the trust of the investigating committee, but, actually, I was giving them just enough. On the other side, I was a terrible friend to John. So, I failed everybody as I see it now. After this challenging investigation, I earned the emotional toughness needed in similar situations to set aside all personal affiliations and focus solely on the safety investigation part.

The bottom line is that personal affiliations in the work context can lure us into being on the wrong side of things. At first, you might think you are doing the right thing, but this can be quite the opposite in fact. In the context of safety investigations, relationships and emotional attachments can easily cloud an investigator's judgement and foresight. Impartiality and placing duty on top of your personal beliefs can be gained through experience. However, this consideration remains on the ethical spectrum of each one of us, so it would be unfair to generalise and point fingers questioning the professionalism of others. The best path for everyone is to search himself/herself where the boundary is and decide maturely to offer services knowing that he/she will be objective and impartial enough. In any other case, the wise thing to do is to abstain, rather than to delay or obstruct the generation of valuable safety recommendations for the whole organisation.

Notes

1. Agosto, Denise. (2016). Why Storytelling Matters: Unveiling the Literacy Benefits of Storytelling. *Children and Libraries*, 14(21), 21–26.
2. Heath, C. & Heath, D. (2010). *Switch: How to Change Things When Change is Hard*. New York: Broadway Books. p. 19.
3. Ajzen, I. (1991). The Theory of Planned Behaviour. *Organisational Behaviour and Human Decision Processes*, 50, 179–211.

Chapter 8

Passenger Experience and Safety Systems

Genovefa Kefalidou

Contents

Complex Mobility Hubs such as airports provide an excellent environment and context for understanding, experiencing and acting upon safety-related issues that manifest within a socio-technical system setting. At the same time, they facilitate, and often demand, a variety of different interactions among people from different backgrounds, roles, goals and expertise. For example, passengers (role) that travel for leisure (goal) must interact with airline personnel (role) to complete a predefined and compulsory check-in (goal) within the expected service cycle of airport travel. While both passengers and airline or airport personnel have roles, goals, backgrounds and expertise, their priorities for demonstrating and executing these can be very different; the same applies for their experiences and interactions as well. Passengers who, for example, look forward to starting their holidays, may experience the queueing at security with impatience or even dissatisfaction although they are aware that there are certain standards to be applied there for their own safety. Similarly, a security staff whose goal is to ensure and execute safety standards (e.g. from screening passengers and their bags to remove potentially dangerous items such as bottles of liquid and pointy objects) interacts with the other actors of the system (e.g. passengers) in such a way to fulfil their operational priorities and goals (e.g. ensuring compliance with airport security and safety standards and movement of queueing flow).

Below, I will unfold one success and one failure ethnographic-oriented story. Both stories refer to safety-related incidents that took place during my work within a project looking at airport travel and particularly on passenger experiences. Part of this project was to introduce technological innovations for airport travel as a response to passenger needs. As part of my role as a human factors specialist, I had to travel to a number of different airports and capture experiences, observe behaviours and identify emerging needs and issues regarding airport travel. Consequently, my experiential role involved being both a practitioner and a passenger at the same time. This, in effect, has been quite fortunate and insightful. Retrospectively, I realised that being immersed within a particular role (e.g. a practitioner) can make you oblivious of situations where 'hidden' safety risks can occur or can side-track you of associated risks due to contextual pressure and distress. For example, a passenger that runs across the airport to catch a plane under stress may miss or bypass certain safety standards due to being in a hurry, which can lead to further delays to the journey.

Humane Moments While Dealing with Safety Issues

I was travelling with a colleague after having attended a conference abroad with a direct flight to my final destination. While I was a passenger travelling for business, I was also collecting travel experience data as part of piloting a study for the project I was working on at the time. The airport I was flying from in this journey was a medium-sized airport with high numbers of passenger flow, especially during that period. After having had some food to eat, my colleague and I decided to go through the security point as we did not have to drop any hold luggage at that time. As it is known within prior airport surveys,[1] security points at airports are considered to trigger feelings of stress and dissatisfaction to passengers. Reasons for that include crowdedness, stringent processes, which are not necessarily mediated through appropriate space, and lots of queueing time.

By the time we entered the security check area, the waiting queue was piling up and passengers were stacking one behind each other while being instructed by the ground staff to streamline the process. While having scattered security staff across the different queues to shout instructions at the passengers, this, as a process, was also quite stressful and difficult to handle; a lot of background noise impaired the capability of passengers to not only understand but actually hear the instructions. This led people in the queues getting agitated and shuffling around. Furthermore, the lack of openness in space and the crowdedness while passengers were carrying their hand luggage made it even more challenging to handle their personal items; this was causing further delays in the security check process. For example, while security staff were shouting aloud the requirements for screening, passengers kept commenting, some of them disapprovingly, that this was confusing as in their previous airport they did not have the same requirements; I presumed these passengers had a connection flight in the present airport.

In effect, this situation led passengers to express not only dissatisfaction with the process but also delay the movement of the queue as they were trying to unpack and shuffle around the items which they thought they did not have to unpack or take off during this time. A passenger with a child and a laptop at hand were behind me; to my understanding, they were travelling alone. She was in a hurry to unpack her laptop and while she was carrying the baby, did not manage to hold the laptop in place and it fell off her hands, landing on the security check floor. Thankfully, the device did not break as it was checked later after the security check finished; even more importantly, the baby was not dropped, and it was just the laptop that was dropped! I can't help thinking what would have happened if the child was on the floor instead of the laptop. What happens with responsibilities when accidents like these can happen? In such cases, who is responsible for what? Who is the ultimate gatekeeper of safety and security within airports?

Obviously, there were other passengers around too, and the scene which had just happened triggered annoyance and surprise in the surroundings. However, as the incident unfolded before our eyes, the security staff came over to check up on things and ensure that everything was fine, which was a very positive sign. While we were approaching the belt area in security, where higher numbers of security staff were around, I started chatting to the passenger that had the dropping incident to try to understand how this happened; there were limited, yet dedicated, spaces just by the queues where passengers could have unpacked their items. The passenger, although she expressed her appreciation for the presence of the security staff and their interest to help when the incident happened, also mentioned that she did not notice the small spaces to unpack her items. She said that even if she had spotted them, she would not use them as they were in a hurry to go through security to catch their flight. On top of that, since she was travelling alone and had to deal with a toddler, unpacking items from her luggage at the same time would have made it even more difficult.

Her feelings were a mixture of both gratitude for the fact that something worse had not happened but also frustration as she was blaming the security and the airport for the stringent processes, which, according to the passenger, posed some unexpected safety issues. This made me think that, indeed, these kinds of incidents can heavily skew the positive experiences a passenger can have within airports, and, eventually, affect their overall travel experience. Furthermore, I came to realise that safety issues seem to be not only about compliance with certain protocols and failure of systems or human error but can also be very much about handling unexpectedness and uncertainty, especially during chaotic and stressful contexts (e.g. rushing through overcrowded security areas with certain constraints).

As we were chatting about this, I decided to include the security staff in the discussion; to my happy surprise, they responded very well and engaged with the topic. Despite the awkwardness created by the previous incident, the chatty behaviour of the security personnel was very positive, they checked again that everything was fine and they seemed empathetic with the passenger. Not surprisingly, they also emphasised to the passenger the presence of the side spaces for unpacking luggage. However, interestingly as well, they highlighted the weaknesses and constraints these limited spaces have. For example, security spaces are often very limited due to space legacy issues that each airport has (e.g. not being feasible to expand the space as other spaces are already built and used around it since a long time ago). Furthermore, they seemed to acknowledge the importance of space design but also making all passengers aware of what packing/unpacking strategies they can employ when being in overcrowded critical spaces and airport touchpoints such as security.

Both the other passenger and I, as a specialist, raised with the security staff the issue of inconsistencies across the airports regarding security protocols (e.g. what needs to be unpacked during the security check process). The passenger explained how this adds confusion and load to any passenger, especially when travelling alone or with vulnerable people. Particularly, in her case, she argued that even though she was aware that she had to take her laptop off her backpack, she couldn't carry it alongside with carrying a toddler for long queueing periods. Interestingly, the staff initially kept emphasising that all airports have the same requirements regarding what gets unpacked during the security process. However, both the passenger and I insisted that empirically speaking this was not the case, and we gave examples.

As we further chatted with the security staff, by the time we reached to the end of the process and started packing our items back into our bags I felt we had developed an interesting rapport and empathy between them and ourselves, a situation which does not normally take place in security touchpoints at airports. The fact that a constructive dialogue among us took place helped the passenger to feel acknowledged and to get to ask questions about the process. Furthermore, I felt the whole scene strongly highlighted the importance of 'humane' (i.e. friendly) interactions even within critical security contexts as the passenger, and I certainly thought we did well to open up a discussion with the respective security staff about the process and the space as well as the 'hidden' safety risks associated to them. The reason why I consider this as a success is that at the end of the day, the passenger affected by the incident said she felt better. Rapport and understanding by the security staff improved the passenger experience despite any safety risks; indeed, the passenger used the smiley feedback buttons at the end of the security process to provide positive feedback.

Disconnecting Passenger Experience from Safety

This time I was travelling solo, coming back from a project meeting which ended up being quite challenging as I had a nasty incident with my back. Therefore, I had to adjust my return trip to be arranged with mobility support (i.e. wheelchair). This was for me not only unexpected but stressful as it had never occurred to me before. At the same time, I was curious to see how this would pan out as my research focused on passenger experience at airports. At this point, it is important to note that my return flight was not direct; I had to connect to another major airport before reaching my final destination.

Changing the status of my return ticket was supposed to be straightforward. I went to the airline desk as instructed and waited there for the desk to open. Part of changing the return ticket was also to declare my hand luggage and backpack as hold luggage to avoid additional burden during my flight. The airline staff performed the amendments and confirmed with me that I was assigned mobility support. Someone at my connecting airport would come and pick me up from the plane to take me to the gate to catch my connecting flight towards my final destination. Upon a request to the check-in staff, I even confirmed the arrangements by checking the information on the staff's terminal computer myself.

In the meanwhile, the airline staff forwarded me to the mobility support service of the airport, which was an independent entity, and they advised me to queue up until their offices opened. Since I could not move due to the problem with my back, I could not do otherwise. I waited until the office opened and I arranged the pickup to take me to the security point through the mobility support route using a wheelchair. Even though the mobility support staff were late picking me up, we rushed through the way to the gate following pathway and process shortcuts according to the protocol. During this time, I must admit that the process was not transparent to me at all, even though I knew in principle what the protocol includes. I was only in a position to guess what was going on and how I would reach the gate in time.

From a passenger's point of view, this situation can be quite stressful, especially if you have not experienced anything similar before and if you are in an unfamiliar environment (e.g. an airport you have not used often). From a practitioner's perspective, it is both interesting and disconcerting, especially when you realise that, on the one hand, protocols are there to safeguard certain safety procedures, but in practice, they contribute only one part to the whole picture of travel experience. For example, knowing that we would follow certain shortcuts and join different queues in the system to reach the gate did not necessarily facilitate peace of mind to me. In reality, I experienced rushing as mobility support were late and we had to hurry to be at the gate in time for boarding as well as uncertainty (e.g. whether everything would be in place for me to have a safe boarding and flight and whether I would catch the flight). Of course, someone would argue that catching the flight was guaranteed in my case as People with Reduced Mobility (PRMs) always board first. So, in theory, at least, the plane would wait for me to board.

The protocol says that PRMs board first on the plane, accompanied by appropriate staff, and they disembark last. After landing, they must wait again until appropriate mobility support staff comes at the PRM's seat on the plane to pick them up. All in all, formal mobility support includes helping PRMs' journey through their departure airport, boarding the aircraft and during the flight, disembarking the aircraft, transferring between flights and travelling through their destination airport.[2] Now, back to the story. As we were rushing to go through security and go across the airport to reach the flight gate, we were following shortcuts (e.g. passing through random doors that seemed to be off the normal passenger path) and banging on doors as the wheelchair manoeuvred around. I must admit that this did not impress me much, and, certainly, I could see safety hazards in the whole process. I did not comment on that at that time, but I kept asking about what was going on. The answer I was getting was that we were slightly late and we were getting to the gate as soon as possible.

Once arrived at the gate, the boarding process started with me being boarded on the plane and explicitly instructed about the protocol and the importance of not leaving my seat once the plane landed at the final destination. I was reminded again that I had to wait on the plane and at my seat for the mobility support from the connecting flight airport to come and get me. At that time and having had experienced the rush to get on the plane in the first place, I was a bit insecure about how the rest of the journey would pan out; I had only 50 minutes to reach the gate for the connecting flight. However, I tried to stay positive and think that everything would be fine. The flight was smooth, quite pleasant and arrived on time at the other airport. I followed the protocol and waited until everyone disembarked. By that time, I was already getting agitated about whether I would catch my connecting flight as disembarking took a long time, and I knew that the airport was large and it would take a while to get to the gate. Despite these thoughts, I tried to remain positive and focus on and trust the protocol according to which the mobility support people would have been notified and would be coming to my seat to pick me up.

Seven minutes passed after all passengers bar myself had disembarked. I called the flight attendant to ask about what was happening as no one had come to pick me up. The flight attendant, who by the way was very friendly, responded to me in a panic that I would miss my flight if I did not disembark immediately. At that moment, I got totally confused as the staff themselves contradicted the protocol and the instructions they had already given me. I calmingly reminded her that according to the protocol, I had to remain at my seat until the right staff came to pick me up. At this point, she thought to have a look at the corridor outside the plane to check whether there was someone there waiting for me. And yes, that was the case! The mobility support staff was outside the plane in the corridor waiting for me all that time to disembark the aircraft contrary to what the protocol dictates. Needless to say, when I realised this, apart from becoming stressed, I got angry as well; however, I did not demonstrate any extreme behaviour.

The mobility support staff placed me on a mobility cart to go across the airport so that we could reach the connecting flight gate, which, by the way, was located at the opposite end of the airport. The staff checked their airport mobile application to confirm the right number for my gate, and once this was confirmed, we set off. The staff was the same stressed as me which led him to drive the cart very fast across the airport, putting in danger not only us but also the other passengers that were strolling around and navigating through the airport. I got scared.

At some point the staff lost his way and had to consult the mobile application again, only to realise then that we were heading in the wrong direction and the wrong gate; apparently, there was a gate number change in the meanwhile. My stress levels had risen again. He turned around and sped up to the other direction, still driving the cart dangerously within the airport. After ten minutes or so, we arrived at the correct gate, but I had missed the flight. I remember the time was 21:00, and it was the time that the mobility support staff's shift ended! I was stressed and alone with reduced mobility within a huge airport!

The staff told me that he was sorry about me missing the flight, but he could not do anything else at that point. I responded by asking what I was supposed to be doing alone on a wheelchair there. Even worse, the wheelchair had to be given back before he checked out of his shift. Hence, he returned the wheelchair while I was sitting at the gate waiting for him to come back. Then he helped me go to the customer/passenger service to see whether there was a hotel to stay overnight. In the meanwhile, the staff found two passengers that they were returning from a gig and happened to miss the same flight. He asked them to take care of me as I had mobility problems, and he left … To stick to the point of the story, at the end of the day, I was left alone with two strangers to try and find a hotel to stay overnight and get the next flight in the morning.

This story aims to highlight a couple of safety-related points. First, the safety issues concerning airport navigation, especially when in a hurry to catch a flight. In my case, the use of a cart vehicle to transfer PRMs to the gate safely and with comfort did not actually guarantee a positive, safe experience neither for me nor for the other passengers in the airport. Navigation when under stress and in a hurry creates multiple opportunities for accidents, and this must be taken into consideration when formulating safety procedures that apply to passengers' movement and not only to airport and airline processes. Passengers constitute one of the key stakeholders within an airport travel service, and their safety should not be contained only within the realms of a security or a check-in procedure. Second, the importance of streamlining communication channels and information exchange across different airports and airline companies but also following the existing protocols (e.g. PRM safe movement) throughout the travel lifetime. What this story demonstrated was that although protocols might be in place and, at least theoretically, communication channels and associated technologies exist, these are not always synchronised to deliver the optimal and safe travel experience service to passengers.

Retrospectively thinking, as a practitioner and a passenger at the same time, I could not have handled any aspects of the above stories differently, given the contextual constraints in each case. What I have experienced and observed was that safety is multifactorial and airport travel has 'hidden' safety issues. In my opinion, we should view and apply safety within a broader context and holistically encompass all associated stakeholder services and not only systems. Also, we must approach safety from an experiential point of view rather than as a discrete desirable outcome. Passenger experience has a flow, is dynamic and not discrete, and, in the same way, safety flows and depends on experiences and constraints (e.g. as in the first story with the limited spaces for unpacking items).

Furthermore, within the airport travel context, complex systems transform into complex services. Thus, there is a need for a multi-sided approach to interactions that guards safety. Safety by experience and democratisation of safety-related experiences between all airport-related stakeholders (e.g. ground airport and airline operators, flight attendants, passengers) could offer new perspectives in understanding and applying safety principles in complex systems. Airports as complex mobility ecosystems do not only consist of engineering systems in the traditional sense. They rather constitute an amalgamation of different types of services and technologies, some of which are not integrated with each other; however, all contribute to what we call passenger experience (PAX).

Moreover, a lot of the new technologies adopted within airports are at higher levels of automation, which transforms the contexts and associated mitigations for their systemic failures. While within complex systems human error is associated with safety failures and non-compliance, human experience and behaviour during travel can also offer affordances for ameliorating negative impacts. Instead of 'losing' this safety-related intelligence, we should capitalise on it and formulate it in manners that it can be meaningfully embedded within appropriate 'live' protocols that 'collate' to a safety library of experiences. This way, both technically 'novice' individuals (e.g.

passengers) and technically 'experts' (e.g. security ground operators or flight attendants) could synergise their experiences and offer a more holistic sense-making approach and understanding and application of safety in travel.

We must understand that Mobility Hubs are living ecosystems of increased complexity and experiential interactions. Therefore, we must achieve a better understanding and inclusion of the 'experience' and 'interaction' factors within the traditional models and approaches to system safety and safety engineering. While, for example, functional hazard analyses are absolutely critical for understanding and designing a system, they focus on functions as opposed to interactions or experiences of interactions. At the same time, human error analyses often focus on performance-based interaction approaches as opposed to experiences; this can make it challenging to understand the safety context from within a service perspective. With advances in automation and the Internet of Things (IoT), more complex systems are being transformed to complex services, thus demanding re-approaching system safety from a service experiential context as well.

Notes

1. https://www.sita.aero/resources/type/surveys-reports/passenger-it-trends-survey-2015
2. https://www.caa.co.uk/passengers/prm/passengers-with-disabilities-and-reduced-mobility/

Chapter 9

Safety Numbers and Safety Differently

Keith Johnson

Contents

The Obsession with Safety Figures

When I first started working in the Occupational Health and Safety (OHS) field back in 2006, I was a little naive in what good looked like for safety or what it was that showed you that you were successful at OHS. What I had observed before becoming involved in that field was that OHS people seemed very driven by paperwork, blamed workers for accidents and had a strong desire not to have injuries or cover them up. These people didn't seem to get out of the office a lot; this motivated me to want to get into OHS so that in some way, I could change the prevailing thinking and perceptions.

On my first journey in safety, I was shown and demonstrated that it was all about frequency rates, be that the Lost Time Injury Frequency Rate (LTIFR), the Total Recordable Injury Frequency Rate (TRIFR) etc. In essence, it was all about lagging indicators, the ones that count the unwanted outcomes. I realised I had to look backwards to understand how I could go forward. I needed to wait for an incident to occur or something less than favourable to happen, then understand why it happened and finally look for remedies so it would not happen again.

After about two years as a safety advisor, I was leaving a 'blue-chip' miner to chase other adventures. While I walked out the door, I spoke with the safety manager of the specific company. He informed me that to be successful in safety, I needed to have a good TRIFR; more importantly, this is what I would be judged on. If the rate was high (let's say more than five [5] based on the industry standards at that time), then I wasn't doing well. If I strived for a low number, that was good, and if I hit zero, then wow! In the latter case, apparently, I was a safety genius, my site was safe, my people were safe, we had an excellent safety culture and I had delivered the benchmark of safety excellence; nothing further to see here! Unfortunately, or not, this theory about chasing a low TRIFR in today's world, as research and practice suggest, is misguided.

Nevertheless, I took this ill-advised theory to my next role as a safety and compliance superintendent at some mines in New Zealand. On day one of work, the first thing I chased down was wanting to know what the TRIFR was. Second, I wanted to know what the record was for the most days the site had gone without a recordable injury (i.e. the cases where the site had sent someone to the doctor and the worker had received medical treatment). Upon being advised of this, I set about placing a large board at the front of the mines which displayed several statistics such as the number of recordable injuries for the year, total days since our last recordable injury, what the site record was for the most days without a recordable injury etc. This was common practice at that time and, occasionally, at the time of writing this chapter in 2020. I placed the standard mantra of 'Zero Harm' at the base of the board and then I religiously tracked the board and ensured the stats were changed every day. I was talking feverously to the crews about what the board meant. I even appointed them as custodians of the board because they were the masters of their destiny which determined whether the sign looked 'good' or 'bad'. I guilted them, in essence!

My passion for tracking a good TRIFR became my obsession over time. I undertook several initiatives, albeit misguided, to ensure that my good TRIFR remained on track. These initiatives were based on what I had observed or had been shown by other so-called 'safety professionals', had read in resource documents of relevance at the time and what general common practice within the industry suggested back then. I am sharing with you some of those initiatives below.

■ Initiative One: minimise the injury. This entailed trying to manage the injury on-site, whereby a trained first aider would review the injury and try to patch it in-house if possible. This would negate any medical treatment following the respective definitions in the company.

■ If Initiative One didn't work, then Initiative Two: off to the medical centre with the injured worker in tow, where I would unceremoniously push my way into the doctor's room with the worker. This initiative entailed three types of actions, as applicable to each situation.

 – Action Type 1: while the doctor was doing the examination, I would annoyingly repeat to the doctor that the worker could possibly take Panadol instead of Panadeine Forte; the latter medicine would trigger a recordable injury, while the former one would not! Clearly selfish on my behalf, but who cares about the worker; It's all about my TRIFR, right?

 – Action Type 2: I would reinforce with the doctor that the company had suitable duties and I could get the worker to do some really 'meaningful' work such as re-reading/regurgitating some Standard Operations Procedures (SOPs). Such tasks or other similarly suitable duties, whether done at work or home, meant the injury was a Restricted Work Case (RWC) and not a Lost Time Injury (LTI). After all, an RWC looked better on paper for reporting purposes!

 – Action Type 3: if the doctor was on the side of the worker and gave an outcome that impacted the TRIFR, there was the option of 'Doctor shopping'. This is the case whereby one takes the worker to another doctor in the hope the new doctor overrules the old doctor's restrictions (i.e. 'nothing to see here'), and the TRIFR is not impacted!

■ Initiative Three: blame the worker! As soon as a worker presented with an injury, one could look for the reasons the worker was 'at fault', failed to follow a rule, was negligent, didn't have the companies interests at heart etc. Often, the first point of call in blaming the worker was sourcing statements from those present at the event or after the fact. If a worker's compensation claim was lodged, a dispute could be the initial default mechanism. This would possibly engage private investigators to catch the worker being in contravention to what he/she suggested their injury was. Such a battle continues until some form of arbitration/compulsory conference is undertaken where the only winner is the law firm!

■ Initiative Four: fudge the figures. If all else fails and one needs to hit a specific target, consideration can be given to fudging the numbers or manipulating the statistics. This is a simple process which might involve (1) entering a false or misleading lag indicator into the reporting system, (2) simply saying to yourself 'Do I really need to report this incident/statistic?' or (3) justifying to oneself why an incident should be downgraded (e.g. from an LTI to a First Aid Case) to avoid impacting the stats. If one thinks this is a moral or ethical issue, then consider that fabricated figures potentially get presented to boards and executive leaders consistently. These persons have a sense of comfort with those figures due to the sense of relief that comes with a target being achieved or the site/project preserving its bonus.

When remembering my TRIFR-related strategies mentioned above, I also recall the reactions from the workers in my previous workplaces, where a lack of trust in management was an everyday reality. There was an element of fear at times in reporting incidents, which often resulted in interventions from the union and simply became a time-consuming process. But there was more to the issue at hand that impacted the workforce. As I relate to Initiatives One and Two, the workforce did not appreciate this process, but they were bound by it, albeit as poor practice as it was. I can imagine them thinking silently something like 'they know better...', 'it must be for the good of the company and all of us...' etc. However, the workforce does not consist of fools, and staff can see through the company jargon, business practice nonsense and misguided company injury treatment ethos. The fact they do not speak up does not mean that employees do not understand what is happening around them.

Actually, the workforce had realised that injury management was all about the statistics, hitting a frequency rate number and making the company look good. God help you if you were the worker that threatened the stability of a site work-free injury record! Furthermore, employees were manipulated and psychologically massaged not to impact company safety stats. If they did, negativity towards them could prevail such as not transferring from casual to permanent employment, not getting more shifts, not having a contract renewed or simply being moved on. I must admit that the workforce was absolutely sceptical of my role and agenda. I think I was perceived as a company man where the worker was not to be trusted at all. I am quite sure I was seen as a 'yes' man that towed the company line and never departed from the initial philosophy of the worker is always wrong, the staff is at fault, or it is only about human errors at the sharp end! Essentially, I had no moral compass, my ethics had gone out the door and I was brainwashed into thinking that safety culture was determined by safety statistics!

What were the key drivers to my thinking mentioned above? First, I had been given poor advice from those that had not really done the research but claimed they knew the job. Second, my lack of knowledge in a sector and field that I was not familiar with, and, third, a passion for pleasing. Corporate officers wanted low TRIFRs, and because I didn't know any better, I followed their misguided lead! However, in Keith's view of the world, the mantra of Zero Harm is lousy practice, aspirational, virtually impossible to achieve. Actually, Zero Harm makes no sense to the people on the ground and senior managers are the only ones that seem to think they understand it; essentially, Zero Harm makes Zero Sense! If someone has Zero Harm in their title, such as Zero Harm Manager, and someone has an injury at the workplace, the Zero Harm Manager should resign because he/she failed in the role in not achieving Zero Harm!

The Shift to Real Safety

There was a moment when the penny dropped and I awoke from the slumber of pursuing Zero Harm and low TRIFRs. The eye-opening happened when I went to a safety conference in Hobart in 2016 and I heard a presenter, G Smith, opening the conference by saying the following sentences, *'Good Morning, I am going to make a statement that shall challenge your thinking...There is no mathematical validity in a Total Recordable Injury Frequency Rate. Nor does it tell you anything about the effectiveness of your Occupational Health and Safety Management System'*. From that point on, I conducted a lot of research on topics such as why a low TRIFR makes no sense, the fallibility of humans in the workplace, safety psychology and science. This inquiry has led me on a new path in the OHS space and culminated in giving my own presentation at the Regional Safety Conference in Newcastle, New South Wales, in 2019. The topic of my talk was 'How Do You Prove Safety Works', which I explain in this second part of my chapter.

My initial research went down the path of understanding what Due Diligence was, what it meant, how it was defined as per the legislation and in lay terms and how it could be achieved. I reviewed what Section 27 of the Harmonised Safety legislation in Australia stated as it related to the Duties of Officers and some practical examples and/or other scenarios where employers hadn't applied proper due diligence. Amongst others, these examples included (1) an Adelaide trucking boss who was jailed for 12 years over a driver's death caused by faulty brakes where the trucking boss knew of the faulty brakes,[1] and (2) the quarry manager jailed over safety breaches that led to the fatality of a 21-year-old worker in a running conveyor belt.[2]

I took these as instances of when due diligence hadn't been readily applied by an employer and what the ramifications could be when proper due diligence is not enacted. Similarly, I started to look at what due diligence looked like from an organisational perspective and I made the following conclusions:

- The organisation knows the hazards in the area of its responsibility.
- The organisation needs proper systems to manage hazards to an acceptable level.
- The organisation ensures adequate supervision to check whether those systems are implemented and effective.
- Individuals comply with the systems (e.g. intended use, procedures).
- The organisation knows how the hazards must be controlled, as foreseen by the systems in place.
- The organisation understands how effectively the hazards are actually controlled (i.e. residual risks).

Moreover, I reviewed why organisations, although they have been trying to prove Due Diligence, finally fail. Such was the finding of the coroner's inquest into the death of Cameron Cole; Cameron was unloading metal frames from a truck trailer, and a metal frame fell and crushed him. In the findings at the inquest, the commentary was made about the use of risk management tools (e.g. risk assessment documents and risk registry). In 2015, the Office of the State Coroner suggested that *'The identification, elimination or minimisation of risks through risk management processes may lead to the production of a suite of documentation that will pass audit requirements. However, the evidence at this inquest suggests that workers in the field may find such documents hard to comprehend and of limited relevance to their daily activities'.*[3] Therefore, in this case, we have an employer who was trying to demonstrate due diligence by conducting risk assessments, but the end-users did not find them useful or did not understand them.

Furthermore, I considered evidence of how Piper Alpha managed injuries on their oil rig before the fateful explosion in 1988. Piper Alpha's injury prevention methodology was to fly the injured persons off the oil rig and back to an expensive Aberdeen hotel, known as the 'Royal', here they were 'treated like a king' and assigned light duties. Everything was about negating an LTI and avoiding an impact on the company's bonus of a quarter of a million pounds! It was well known on Piper Alpha that the statistics were fiddled, and arguments were aplenty in safety meetings about the classification of LTIs or why LTIs were not reported. Indeed, the safety department was no longer called as such; they were better known as 'the bluff department'.[4] Additionally, I reviewed the 2010 disaster of the Deepwater Horizon Macondo Oil Rig where the oil rig exploded and sank, killing 11 people and seriously injuring a further 16; also, five million barrels of oil were released into the Gulf of Mexico.[5] On that fateful day, the Macondo project celebrated seven years LTI Free, but the findings suggested that Macondo had a focus on personal safety instead of process safety.[5]

In addition to the above, I reviewed the Esso Explosion in Melbourne, Victoria, in 1998, where two workers were killed and eight staff members were wounded together with a loss of gas supply in Melbourne for several weeks. The Royal Commission severely criticised Esso's safety management system by stating that *'the OHSMS comprised a complex management system...it was repetitive, circular, and much of its language was impenetrable'.*[6] Ironically, Esso's safety performance the previous year, in 1997, went by without a lost-time injury; they even won an industry award for this performance.[6] The Commission further stated *'The lack of focus on process issues is a matter of grave concern. To put it bluntly, Esso's focus on lost time injury rates distorted its safety effort and distracted the company's attention from the management of major hazards'.*[6]

I also studied what Diane Vaughan referred to as 'Normalisation of deviance' as it related to the explosions of the Space Shuttle Columbia.[7] NASA was aware that the foam protection plates that covered the outer shell of the shuttle came off during take-off and it had occurred on other take-offs before the explosion on 1 February 2003. However, because the foam had not impacted the shuttle nor caused damage during the previous occurrences, NASA seemingly accepted the risk and normalised the deviance.

Consequently, the time to reflect on my practice came. I pondered on how the safety fraternity had adopted this massive fraud, where we fudged the figures, gave SOPs to workers in hospital beds to negate the LTI or simply didn't tolerate the statistic and thus manipulated it to our advantage. This wasn't really a legal issue, but more of an ethical and moral problem and the whole scenario had nothing to with safety anymore, it was all about the numbers and whether we look good. As Sidney Dekker stated so eloquently, *'There's only one number we care about...It's my LGI...How's my LGI...How's my Look*

Good Index![8] I looked at all kinds of reporting (e.g. reporting of accidents or reports to boards) we were maintaining and consulted the response from the Royal Commission to the Pike River Coal Mine Tragedy. Their comments included how the Board of Pike River received some health and safety information in board reports based on statistics and time lost through accidents but *'the information was not much help in assessing the risks of a catastrophic event faced by high hazard industries'*.[9] Similarly, the Commission went on to state that the Board appeared to have received no information proving the effectiveness of crucial systems such as gas monitoring and ventilation, both of which are vitally important in underground coal mines.

Further examination of our reporting practice led me to think that lead indicators should give me some insights into whether the Occupational Health and Safety Management System (OHSMS) is effective or not. However, I often see that lead indicators appear as nothing more than a measure of activity; achieving 100% compliance creates an illusion of safety! Because indicators are a measure of activity, they tell us that things have been done. However, they do not tell us anything about the quality or effectiveness of the activity and the influence of the activity on safety, and they do not actually contribute to proving the effectiveness of our critical systems. Unfortunately, safety reporting plays a significant role in helping the illusion of safety...or...the 'Safety Paradox'. The safety paradox is the gap between the system as written versus the system as imagined versus the system in practice; briefly, these gaps are termed as work as imagined versus work as done. We have this philosophy and belief that in our systems doing paperwork and completing activities means our risks are controlled; however, this is the illusion of safety or better described as being 'Paper Safe'.[10] We fill in a lot of paperwork, but unfortunately, this doesn't tell us a whole lot about whether risks are controlled!

So now that I had gone and reviewed all of this information, what was it that I needed to do next? I adopted a whole new philosophy. First, I questioned myself and the senior leadership team about what information we get as managers that proves the effectiveness of our critical systems. My thoughts are that if all we are receiving as a management team is information about injury rates, then we are not exercising due diligence. Similarly, I don't accept low injury rates means the organisational risks are being managed nor that a green traffic light report suggests critical controls are managed or are effective.

I have changed my focus around the things that go bad or didn't go to plan and adopted Erik Hollnagel's philosophy[11] of much more goes right than goes wrong. This suggests that for every 10,000 events, 9,999 times they go right and only once they don't go to plan; yet, which number do we focus on? I have also tried to adopt Sidney Dekker's Safety Differently approach,[8] where we see people being the solution and not the problem. The workers have the answers because they do the job; it's the presence of positive capacities (i.e. what went right) and not the absence of negatives (i.e. reduction in incidents/incident reporting). Finally, it's an ethical responsibility of managers to lead from the front and by example and being transparent in their approach.

I also looked at how one can simplify the rules and take away the 'safety clutter'. One example and philosophy that seemed to work was Hans Monderman's 'Shared Space' intersection design in Drachten, the Netherlands.[12] The intersection had a history of accidents and adverse outcomes, so the traffic engineer Hans Monderman said to take out all the signs, lights and controls. Because there were no rules for the intersection, spontaneously drivers slowed down to the lowest common denominator. It's about trying to find a happy medium ranging between full autonomy like Monderman's intersection to an environment wholly clogged with rules!

In closing, what did I take away from my research and experience? I continually ask myself 'How do you prove safety works?' because once we are in court because of a workplace accident, our health and safety management system is deemed futile. I always think about the safety paradox we work in, where there is a disconnect between work as imagined and work as done. I see TRIFR as a meaningless measure that adds to the illusion of things being in control! I reflect on incidents like Macondo…Esso…Texas Oil Refinery…and how all had a declining injury rate and then they had catastrophic events! Have we been fooled?

From what I have shared in both of my stories above, time and experience have shown me that good safety performance has nothing to do with the numbers or the statistics in a frequency rate etc. It is about how you prove safety works and that is all about due diligence. Your due diligence is demonstrated by showing that you have done what you intended to do. For example, you grab a site Safe Work Method Statement (SWMS) and walk onto the site and validate that the workers are following the process. If they are not, then why not? Is it because the SWMS does not work because the workers do not understand it or it is too complex or it was simply a 'tick and flick' exercise? If I relate due diligence to an incident, I shall review the corrective actions from the incident and go out into the paddock and check that they have been implemented, but also determine whether the corrective actions are effective or they have simply created another hazard! If it is an inspection, I will grab the completed inspection and go out on-site and look at what has been inspected to validate it is correct and that the activity of completing the form was not a paper exercise.

If due diligence is real, then you have a management team who are prepared to adopt it and want their staff to feel empowered to own it. I worked for a general manager who was new to safety but was very open to learning. With time he accepted my change in thinking, he agreed when I challenged what was seen as the norm in safety and he endorsed that safety was taking a new direction. That manager realised that when someone has this kind of support, they can achieve almost anything in the safety space as it encourages people to speak up, question the unknown and ask the 'dumb' question. Under such conditions, safety simply sits in the background and becomes unified with all the workplace elements.

Proving that safety works is an ever-evolving process and, to my perception, the regulators are still in a catch-up mode. They tend to live exclusively in the compliance space 'oh but the legislation says…'. Well, guess what! Safety is not always black and white; more often than not, safety is grey and that is because it involves humans with variable performance, their own perceptions and attitudes, creative spirit and whatever distinguishes humanity from technical and fully reliable components. It is true that some safety practitioners still tend to believe that TRIFR, statistics and leading and lagging indicators are the 'go-to' for proving how safety works. I hope that this trend will slide as long as we continue looking for opportunities for improvement and emerging best practice methods in the safety space.

Notes

1. https://www.abc.net.au/news/2015-08-21/trucking-boss-peter-colbert-jailed-for-drivers-death/6714506
2. https://www.abc.net.au/news/2019-05-24/court-quarry-worker-death-sean-scovell-mcg-quarries-qld/11092980
3. https://www.courts.qld.gov.au/__data/assets/pdf_file/0008/436589/cif-cole-cb-20150911.pdf
4. Beck, M. & Drennan, L. T. (2000). Offshore risk management: Myths, Worker experiences and reality. *Paper presented at the Qualitative Evidence-based Practice Conference, Coventry University, 15–17 May 2000.* http://www.leeds.ac.uk/educol/documents/00001394.htm

5. McNutt, M. K., Camilli, R., Crone, T. J., Guthrie, G. D., Hsieh, P. A., Ryerson, T. B., Savas, O. & Shaffer, F. (2012). Review of flow rate estimates of the Deepwater Horizon oil spill. *Proceedings of the National Academy of Sciences*, 109 (50), 20260–20267. DOI: 10.1073/pnas.1112139108

6. http://www.futuremedia.com.au/docs/Lessons%20from%20Longford%20by%20Hopkins.PDF

7. Vaughan, D. (1996). *The Challenger Launch Decision: Risky Technology, Culture, and Deviance at NASA.* Chicago: University of Chicago Press.

8. Dekker, S. (2015). *Safety Differently.* Boca Raton: CRC Press.

9. https://pikeriver.royalcommission.govt.nz/Final-Report

10. Smith, G. (2018). *Paper Safe: The Triumph of Bureaucracy in Safety Management.*

11. Hollnagel, E. (2014). *Safety-I and Safety-II: The Past and Future of Safety Management.* Farnham: Ashgate.

12. https://www.nytimes.com/2005/01/22/world/europe/a-path-to-road-safety-with-no-signposts.html

Chapter 10

Infrastructure Projects as Complex Socio-Technical Systems

Maria Mikela Chatzimichailidou

Contents

At the time that this chapter is being written, I am a Team Leader and Principal Engineer in the System Engineering, Integration and Assurance department of a multinational engineering and design firm. I switched to this sector from academia two years ago in an attempt to get outside of my comfort zone and become more familiar with the corporate world. Since both of my stories regard rail safety, let me explain here the overall framework governing risk management in this sector.

Safety risks are controlled through a risk management framework known as the Common Safety Method for Risk Evaluation and Assessment[1] (CSM-RA), a European Union (EU) Regulation introduced by the European Commission to provide a common process for risk analysis and evaluation across EU member states. According to the CSM-RA, there is a series of steps safety engineers need to follow to deliver safe designs and operational systems, without, however, being prescriptive on the techniques and tools to be used. One of the most important steps of the CSM-RA process is the performance of a Hazard Identification (HAZID) workshop, which, to be reliable and successful, must be based on a robust system definition. In the System Definition stage, the Safety Assurance team must define the system under assessment (i.e. objective, functions, boundary and interfaces, environment, existing safety measures) and state any assumptions that determine the limits for the risk assessment.

Almost Missed the Forest for the Trees

One of my first tasks as a new starter and a trainee was to support the lead safety engineer in producing the safety case for the installation of safety screens along a platform for a high-profile rail project in the United Kingdom (UK). As a matter of explanation, in the majority of the cases, a trainee acts as a scribe and is not engaged in the development of the safety case. Safety screens are physical barriers installed along with other parts of a platform above the standing surface. Their use helps prevent people from committing suicide, accidentally getting in contact with moving trains and entering tunnels for any reason; also, screens do not allow objects to fall on the tracks. Additional benefits include the improvement of climate control within the station, with a positive economic impact too, and protection from adverse weather conditions when used in the open air. According to the safety case requirements of the contract, the Contractor had to deliver to the Client a comparative study of different configurations of safety screens mapped onto a list of criteria such as design life, maintenance requirements, conductivity, fire resistance, compliance with UK rail standards, maximum panel sizes, availability and cost.

As a first step, the Contractor's System Assurance team, to which I belonged, defined the boundaries of the system under consideration, which was the actual platform. The Contractor presented this in a report including the boundaries of and interfaces within the platform so that to consider later the mitigation of known risks within the defined system. The general arrangement of the platform was provided by the Client and included by us in this report. Based on this generic design, the Contractor identified the system, the system interfaces (e.g. platform and earthworks) and the people affected by the system (e.g. passengers, staff) as well as any assumptions that would be useful for the later stages of the project, usually related to platform refurbishment and upgrade. Possible interfaces with other projects and environmental factors were not examined because the platform was located in a confined area.

After the system definition report received the Client's typical approval, which is usually the case since the system of interest is described in the contract signed by all parties involved, a HAZID workshop with representatives from the Client and the Contractor was held. At the end of the HAZID workshop, the Client and the Contractor collectively identified roughly 40 risks, with 15 of them being labelled as broadly acceptable. According to the common mandatory European risk management process for the rail industry, the 'broadly acceptable' classification applies to insignificant or negligible risks. This could be because any of the respective hazards is so unlikely to arise that there is no need to install measures to control the risk it creates (i.e. very low-frequency event) or where there is a credible failure mode, but the consequences are negligible (i.e. high-frequency/very low-severity event). An example of a very low-frequency event is a meteorite impact, and an example of a high-frequency, very low-severity event is a 'paper cut'.

Here, someone may wonder: How does someone decide that? Are there reliable and valid criteria? How do you know what you know? The answer is: They just know! The HAZID workshops are attended by established engineers across all disciplines (e.g. civils, telecoms, safety), and it is only their knowledge and experience that ensures that all hazards are identified. A reasonable question is: Could this be done differently and are there any external sources to consult with? Generally, the answer is yes, but provided that this specific project was classified by the safety project team (Client and Contractor) as low risk and small-scale, there was no external assessment body (AsBo), as we call it in the railway industry. The AsBo is an independent and competent external or internal individual, organisation or entity that undertakes an investigation to provide a judgement, based on evidence, of the suitability of a system to fulfil its safety requirements. So, with all these questions flying through my head, after having recorded all the hazards that needed monitoring under normal railway operations, the Contractor identified safety measures and requirements and, as usual, prepared the risk register. The latter is a database in the form of a spreadsheet where the Contractor records all the above information and evidence, accompanied by a follow-up report. Our report explained the history of this safety case from the definition of the system to the codes of practice used to provide detailed practical guidance on how to comply with legal safety obligations.

The project wrapped up and the safety engineers moved on with other projects. Altogether, a normal succession of events. The project was left with the Project Manager (PM) only who had some extra paperwork to do before issuing the final reviewed document for the archives of the Client and the Consultant. The PM, a former hands-on safety engineer who had the overview of the project and signed off the reports, picked up on something very important, which surprisingly none of us, the technical experts, raised before. His question was: What could possibly happen in the case of a terror attack or fire where evacuation is mandatory? He contacted the whole project team, i.e. Client and Contractor, directly. The news left the project team totally flabbergasted – but not me for reasons I alluded to previously and I also explain below. I remember the PM, who I was acquainted with, saying: How could the project team miss such a crucial scenario? Why on earth did you consider normal operations alone and the hazards on days of regular railway functioning? There were trust and respect between us; thus, I confessed to him my related concerns during that HAZID workshop, without though elaborating further as I thought it was not the right time for a blame game. By the way, in the UK industry, it is not a common practice to involve the PM in the review of the technical specifications; the main task of this role is to manage the budget, schedule and resources.

Before I continue with my story, it is important to mention my feelings as a trainee systems assurance engineer. I was a person who had just landed in the industry after having spent a decade in academia and hence my confidence had reached historic lows! You see, in academia, I used to treat safety and security as inextricably linked. The approaches I adopted as a research student and later as a professional were systems-theoretic and hence it was the safety process itself that was forcing me to identify scenarios where safety measures could have security implications and vice versa. So, to go back to my story, I must admit that in that HAZID workshop I thought about raising my concerns in relation to security, but at the end, I never did, as I thought I would look somewhat of a fool. What did I know? I was an academic who had just switched to the industry, like a fish out of water. Obviously, I was wrong – I should have spoken up! Undoubtedly, those engineers with the long-standing experience did all the dirty jobs, identifying the biggest bulk of hazards and safety requirements. However, I had a fresh eye and an out of the box thinking that could have helped us reach the maximum of our potential as a Client–Contractor team. It was my fear in combination with some other unfortunate incidents I experienced, where established engineers had resisted change by suppressing and underestimating different opinions and viewpoints. As you would expect, my trust in them lowered while my confidence really started improving.

To resume the story, within a week since the project team got notified by the PM about this omission, because of its significance, our involvement in other projects that we had already started working on was recalled. In turn, all our attention was now concentrated on the neglected scenarios. Within the same week, the Contractor had a workshop with the Client in a desperate attempt to figure out what went wrong. In that workshop, it was discussed that those screens could possibly prevent passengers on the train from evacuating and getting access to the platform. Moreover, the Client brought to the Contractor's attention that apart from the evacuation scenario, the Contractor had been missing some aspects of maintenance as well. That is, the station staff would occasionally need to access rail equipment in the cupboards placed in narrow areas at the end of the platform not accessible to the general public. Usually, station staff are not trained on Personal Track Safety (i.e. safe working practices for rail workers), and, therefore, these areas would need additional safety measures to meet requirements for working areas adjacent to live running tracks. The workshop concluded with the realisation of those two additional observations: hazards introduced by the safety screens in case of evacuation in combination with the possibly conflicting scenario that screens shall fulfil safety specifications to protect station staff working at the two extremities of the platform. The Client went back to their design teams and the supplier responsible for the manufacturing of the safety screens and came up with a specific design option, the risks of which the Contractor should assess.

The Client produced a document to inform all station Contractors in the procurement process, explaining the requirements, materials, details, arrangements, constraints and layouts at each station (i.e. system of interest specifics). According to that report, a 2.5 m high screen should be installed so that station staff are protected from passing trains while being able to access maintenance cupboards. The emergency walkway from the tunnel into the station should remain open, providing a minimum width of 1000 mm between structural elements (e.g. any steelwork, platform edge doors, storage area door) or 1100 mm between any element and a platform edge. The generic screen design considered the train gauge and engineering loadings. It should be properly bonded to meet the applicable Earthing and Bonding requirements. The report also described the process that had led to additional design decisions after the new hazardous scenarios had been identified. The Contractor's responsibility was to use this information and make station-specific designs by considering the additional hazardous scenarios. The Contractor conducted a site-specific as-built assessment of the end of platform areas, including the safe access route, a full spatial survey and an assessment of earthing of each asset type. To support the detailed station-specific design, the Contractor was provided with a set of station mark-up plans showing the linear extent of the screens sourced from as-built and detailed drawings.

As one would expect, following that lifesaving (without any exaggeration!) PM's note, a new contract was signed and we proceeded fast to the follow-up project and an additional HAZID workshop for the new hazardous scenarios. Suitable mitigations in the form of safety requirements and assumptions were developed as part of this process. A further 30 hazardous situations were defined, on top of the 40 identified for the case of normal operations. Also, in this workshop, we reconsidered the size and location of the safety screens and decided to replace those already installed to meet safety requirements related to evacuation and staff operating on-site while trains operated. Along with the Client, the Contractor agreed on a 2.5 m high galvanised steel metal structure and mesh screening bonded to traction earth that was spatially separated from existing known station earthed assets. Although it was not part of the formal process we typically follow in the UK rail industry nor a contractual responsibility, we ran an additional workshop with attendees from the Client, the Contractor and the manufacturer of the safety screens. The purpose was to make sure that we were not missing any other safety-critical scenarios and that we could all sign off all the documents required for the Testing and Commissioning stage.

The PM's observation was one of those 'eureka moments' when you realise that you could have unintentionally contributed to fatalities or injuries because you were just too short-sighted to see beyond day-in-day-out operations. The big question is why we all failed to raise the lack of safety study for the case of an emergency evacuation. I was scared for the reasons I explained above. I did not dare to confront the Client in that very first HAZID workshop. Actually, I never told the Client, as there is this perception that the client is always right. But I did tell the lead safety engineer who encouraged me to speak my mind and actively participate at least in the workshops he chaired. Three months after this case, my line manager signed off my promotion. It was the lead safety engineer and the PM who provided feedback on my performance and potential as a safety engineer.

All things considered, the problem, in my opinion, lies in the mentality of the project team. We performed siloed work and adopted a narrow view of the project. As I reflect on the discussions back then, I remember one of the safety engineers saying something along the lines of 'Security is not my problem; let someone else pick up on that'. Someone else from the steelworks team said, 'We should hit the time targets so let's sign it [i.e. the safety case] off'. Someone else from the telecoms team said, 'Let's request a CCTV system and first response teams will do the rest'. There were so many different disciplines there, but we failed to follow a holistic approach. We were complacent and had this hidden hope or belief that someone else, later down the line, would pick up on any possible omissions.

Since that case, which after all was a successful hazard identification exercise, I adopted the approach that whenever I attend a HAZID workshop, no matter how big or small, how important or seemingly trivial the project is, I will consider hazards that relate to degraded and emergency modes of operations and request them as part of the system safety plan. At least I will know that I tried. As a final suggestion to the reader: allow different voices to be heard. Age does not always go hand in hand with experience, mindfulness and accurate evaluations. Be exposed to versatile work, and, if you do not get the chance, leave some space for others to chime in.

How Common Safety Practice Can Fail the Overall Project Delivery

I was leading the safety discipline for a small project that falls within a high-profile UK-based rail programme. The project I was accountable for was to safely introduce new technologies for airborne noise mitigation in the rail sector. My company (Contractor role) assigned this project to me as my first leading role in the safety discipline. The sponsor of the project (Client role) wanted to design and install a bespoke solution, which could be a mixture of acoustic barriers adopted around Europe, but 'definitely not a replica', as they had clearly stated. This meant that the project team (Client and Contractor) had to start the safety analysis literally from a blank page. At the same time, it was the Contractor's contractual obligation to prepare a safety case according to the CSM-RA framework.

CSM-RA applies to all significant changes that meet the following criteria, which the (conceptual) system that the Contractor is responsible for risk-assessing has to meet to a sufficient and demonstrable extent:

- Failure consequence: credible worst-case scenario in the event of failure of the system under assessment, considering the existence of safety barriers outside the system under assessment.
- Novelty used in implementing the change: this concerns both what is innovative in the railway sector and what is new for the organisation implementing the change.
- Complexity of the change.
- Monitoring: the ability to monitor the implemented change throughout the system's life cycle and intervene appropriately.
- Reversibility: the ability to revert to the system before the change.
- Additionality: assessment of the significance of the change considering all recent safety-related changes to the system under assessment even if these were originally judged as insignificant.

Normally, according to the CSM-RA process, to define the system in the detail and accuracy required by the law, the Client is responsible for providing a preliminary design as a minimum. Following that, what the Contractor and the Client do in the first HAZID workshop is to agree on an exhaustive list of hazards for the defined system and according to the preliminary design. After the hazards are identified in collaboration with the Client, the safety requirements are generated by the Contractor and then presented to the Client in the form of a report so that the Client can produce or take the lead in preparing and approving the detailed design.

As stated previously, the contractual obligations of the Contractor were to produce a safety case according to the above CSM-RA steps, but without the preliminary design and system definition CSM-RA was impossible to be applied on time, and the solution was only in everyone's imagination. Actually, every single person involved in the safety analysis had a completely different perception of the desired solution. Consequently, from the early stages of my introduction to the project, I understood that the safety case I was liable for was not like any other project I had worked on until that time. The project team, with me leading the safety analysis, had to identify hazards for an inexistent design. Both the Client and the Contractor knew the existing circumstances, but they decided to go ahead with the project, as if it was 'business as usual', as they called it. To put it simply and bluntly, to my mind, the Client wanted to be done with the lawsuit and the Contractor wanted the in-time payment.

Soon, I realised that, for the sake of the safe delivery of such an unusual project, it was unavoidable to alter the common practice and postpone the application of the law that regulates the safety of rail operations in the UK. As the safety lead from the Contractor side, and after having decided to put politics aside, my main driver was to eliminate the chances of sacrificing safety, while at the same time I ignored contractual time and budget restrictions. I decided to delay the CSM-RA steps, meaning that instead of chairing a HAZID workshop straight away, I instructed a pre-HAZID exercise. With this preparatory workshop, I also wanted to make sure that our role as a Contractor was clear, acknowledged and approved by the Client. The Contractor was not paid to do this additional job, which was not part of the contract with the Client either. In theory, in the UK rail sector it is the Client's responsibility to define the preliminary design, while the Contractor is accountable for defining the system, mainly based on the preliminary design, and then identifying hazards jointly with the Client, mainly based on the system definition.

The project lasted three months in total. During this time, although I had too little information about the concept design, I was getting prepared for the pre-HAZID workshop, discussing with the Client what disciplines (safety from the Client's side, acoustics, civils, track, maintenance, rolling stock etc. from the Contractor side) should attend the workshop and trying to define its scope and structure as precisely as possible. As explained above, during the pre-HAZID workshop, there was no selected design solution for which the attendees could identify hazards, but only a number of concepts and available products, none of which singly could meet all of the Client requirements, from which, though, a solution with the desired characteristics should be identified. The attendees from the Client and the Contractor side undertook a feasibility study of the available options of barriers for noise reduction and considered design requirements to remedy the risks identified for each of the known options (i.e. available commercial products). The outcome of the workshop along with the preceding and succeeding documents was a summary of the desired characteristics of a bespoke noise barrier solution, which was satisfactory both in terms of client requirements as well as its technical aspects, i.e. acoustic and safety performance. Apparently, we all failed to meet the purpose of the project (i.e. obligation to produce a safety case according to CSM-RA), which was the identification of hazards through the so-called HAZID workshop.

After the above, the Client decided to postpone the project completion date until further notice to the Contractor because we, as a joint Client-Contractor project team, did not manage to deliver a definite set of design requirements for a safe and efficient airborne noise mitigation solution. However, the Contractor's payment was made in full and without any complication, as the unsuccessful project delivery was described by the Client as an unpredictable complication due to unforeseen project complexity and novelty of the technology sought. Since the pre-HAZID workshop was out of scope according to the clauses of the first contract, part of the new contract (i.e. an extension to the original) would refer to the Client's own responsibility for thoroughly defining the design requirements ahead of any formal involvement of the Contractor or HAZID workshop. The Contractor, on the other hand, would have to intervene later in the process to assess the risks of the design proposed by the Client by applying the CSM-RA process in full this time, before going ahead with the final noise barrier solution, as part of the Client's accountability.

Seven months after the above project, I returned to the same programme, working from a different position this time. This second role was to contribute to the execution of the programme's Systems Integration (SI) strategy, which described the technical coordination of the whole system to deliver the end-state railway. The mission of the SI strategy was to deliver an end-state railway that was safe, capable, operable and maintainable, and ready for revenue service. Looking at the big picture, I realised that almost all the projects within this programme (e.g. installation of an overhead catenary system and possible impact on other systems such as the rolling stock architecture) suffered the same fate as the one I was involved seven months ago. There was no clear scope for the production of the safety cases and the system boundaries were not accurately defined. However, despite the uncertainty, again, the Client required the completion of the safety and other technical studies and the Contractors (i.e. different consulting companies than the one I represented) went ahead despite the lack of information from the Client.

I was not pleased to see that the acoustic barriers project was still on hold. In my opinion, that project was a typical example of an overconfident Client and a hurried Contractor. The Client missed to assess or overlooked the technological uniqueness of the project, and the Contractor rushed to follow the CSM-RA process because this is what the Client formally requested in the contract. On top of that, the Contractor did not want to dishearten the Client or stop being in the Client's good graces and hence did not raise any concern. Apparently, the Contractor did not deliver against the key contract request, which was the production of a safety case.

What we learned from a Contractor's perspective was that we should not rush into contracts that are not sensible. Retrospectively, if I were to recommend changes in the way out-of-the-ordinary projects are approached, it would be that the Client should be reasonable in his expectations from the Contractor, and in turn, the Contractor should be upfront about what is feasible, given the existing regulatory framework and the projects needs as well. In a desperate attempt to hit revenue and utilisation targets, especially during times of recession (e.g. the financial crisis 2007–2008) or uncertainty (e.g. Brexit), contractors take up any work available without performing sanity or feasibility checks. All things considered, we should be treating every project as a single case by thinking of its specifics. Complacency and lack of creativity might render us short-sighted to find ways to meet the law requirements. We should not be following the predefined procedures and legislation just for the sake of ticking another box on the mandatory checklist but progress the project idea to a mature stage and then apply the law as required.

In a world where companies are constrained by relatively dated regulations, countries and governments find it almost impossible to sponsor and eventually deliver versatile safety-critical projects, steps should be taken towards more agile and lean governance of programmes of national, technological and economic importance. Maybe part of the solution would be to introduce regulatory frameworks accustomed to the complexity and uniqueness of projects and hence give space for innovation and new approaches to solving complex problems. When projects are 'virgin', the law could allow some space for experimental design solutions. For example, the medical sector is allowed to do so under strict rules and protocols when, for instance, new drugs and treatments are tested. Regulations for the rail sector can adopt a similar approach, mandating in-depth Client–Contractor conversations and long-lasting testing periods when an off-the-shelf solution is impossible to be identified.

Note

1. https://www.era.europa.eu/activities/common-safety-methods_en

Chapter 11

The Two Sides of the Same Coin

Marion Kiely

"I can see clearly now the rain has gone" is a great song written by Johnny Cash and performed by many artists. Sometimes when we are in the thick of something, it is hard to see the wood from the trees; we are so close to a situation that we cannot get perspective. Only with hindsight, distance and time do we see with greater clarity what was really going on. Such is my story, which brought me to great heights of elation and then plunged me deep into deep depths of despair. Contrary to the guidance from the editors, in my chapter, I am sharing a single story about success and failure.

My tendency to look out for and protect others probably came with the territory of being the eldest in my family. The tendency to mind others was necessary during my parents parting of ways, when I cared for my four younger siblings, and seems to have stayed with me throughout my life. Picking up on non-verbals and having a sense for what is being said when nothing is being said, while at times is a gift, has also proved to be a source of much dis-ease while striving to "keep the peace".

For me, Health and Safety was always a logical veering of sorts. No matter what role I was in, if stuff was needed, I let it be known, if approval was needed, I persisted until it was received, if items of importance were put on the long finger, I brought it back on the short one with persistent follow-up. I was willing to get stuck in and do what needed to be done to make life easier in getting the job done. It was often joked that I was the pain in the life of many managers due to my persistency, some even accrediting their receding hairlines to me!

One role I previously held involved introducing and leading a new safety programme to a pharmaceutical site I was working on with the support of the steering committee, a cross-functional group. I relished the opportunity as I felt we could bring about some much-needed positive change. Although we promoted it as a Behavioural Based Safety (BBS) programme, it wasn't a true BBS programme at all. We refused to count how many employees held handrails and all that other stuff we considered to be unimportant and non-value-adding, and we did not focus on whether any behaviour and measure were or were not improving. A year or so into the programme, feedback from a DuPont consultant was that we were: "fixing too many things and not people". I took that feedback as the greatest compliment we could get, we didn't need to fix the people, but the tools, procedures and systems with which they worked. We wanted to find out where people might get hurt on site and do something about it in advance so that we could prevent injuries; that was it. We were very naive in some ways when I look back, but we were bursting with enthusiasm and passion for what we believed was possible.

It was not until I attended a masterclass in human factors and safety with Professor Sidney Dekker sometime later that I realised we had many components of Safety Differently in our programme, but we weren't aware of it at the time. Coming predominantly from the frontline ourselves, the safety committee saw people as the solution rather than the problem; our main aim was to engage them and keep the safety conversation alive. We stood firmly alongside frontline workers as our functions were also in the frontline. We focused on the presence of positives, such as the continuous stream of observations and related conversations, and we communicated and promoted those across the site regularly. What would have previously been deemed a bad news story was turned on its head and sent out as a good news story because someone took the time to tell us where someone else might get hurt and we were now in a position to do something about it. This approach proved challenging for some of the management team as they saw it as a dig towards them because this "imperfect" or unsafe scenario was present on their patch. Nevertheless, our goal was to bring safety to the forefront of everyone's mind as they went around the site and we needed to communicate it regardless of location.

We were stopped by the union in our tracks early in the programme, while we were still in the preparation stages. Some operators had searched online as to what BBS was and they thought we were introducing a punitive programme and had raised concerns with the union representatives. We put together a rough and ready presentation highlighting what our programme was all about: it was voluntary and anonymous. This meant that there were no repercussions or punishment, and, if people on site did not want to be observed, that was absolutely fine. They would be asked in advance for their permission; if they granted it, the observation would take place, if not, it would not. After this presentation, one of the union representatives sent an e-mail to the Human Resources (HR) director stating: "Marion gave a presentation this evening to the site shop stewards on a new initiative to safety called Behavioural Based Safety. I thought the presentation, concept and delivery to be excellent and will be recommending it fully to the technical group. Thorough preparation was obvious, and any questions from our group were dealt with demonstrable expertise. I fully support this initiative and would be happy to volunteer myself as an 'Observer'. There is a huge benefit to everyone's safety from this idea, from multiple perspectives, and major credit due to the team". This was a great endorsement from the outset and what seemed like a barrier to progress was removed. We felt that we were set up for success!

The programme kicked off and everyone on the pilot site, including contractors, had to attend a one-hour compulsory information session on the programme. We recorded an in-house video with employees who spoke of the impact injuries had on their personal lives, intentionally not focusing on the details of the incident itself. We wanted people to get a sense of the effects on home life and plant a seed that it could be one of them next time. We felt that this was more authentic and personal than showing videos of disasters overseas and would have more of an impact. Over 60% of attendees volunteered to become actively involved in the programme, we were really delighted with the uptake and response. Volunteers attended a one-day training course delivered by the steering committee. With the involvement of an organisational psychologist, we made these training days interactive and fun. We wanted to ensure that all employees got to "go home safely", and our slogan was "tóg go bóg é" – Gaelic for "take it handy" (i.e. slowly). Feedback was invited after each session and we listened and tweaked the course based on this feedback as we went.

Observations started coming in and the amount of information received was amazing and highlighted many areas on site that posed injury risks to employees and contractors, some of which were in remote places on the plant that very few people got to see. The process was that observation forms would be dropped into boxes at various places around the site and collected twice weekly. These were then walked down with the site engineer and safety engineer each week. Actions were raised on various items, and what was really important on the back of this was the feedback loop. The person (operator, fitter, lab technician, contractor etc.) who had put the observation in got a number; in my role as a facilitator, I could go back to them if I had questions at any stage or needed clarifications. I never knew the identity of the person being observed, nor did I want to as it was extremely important to maintain anonymity. The identity of the observers was not revealed to anyone; only the facilitator had that information and protection of anonymity was critical for the success of the programme.

From the time that any observation was handed in, there was regular feedback to the observers who were kept in the loop whenever there was a review or update. The steering committee felt that regular communication helped to maintain engagement. It is often the case when an item is raised through the incident reporting system, good work happens in the background by the safety department. However, if the person that raised the incident is not kept in the loop, he/she may perceive that no action was taken and might believe that value was not seen in what they raised. Such a situation can potentially affect engagement and interaction with the system in a negative way. For that reason, maintaining regular communication was central to the programme. The observations kept coming in and we maintained our focus on getting actions closed out to keep the momentum. The site leader was very supportive of the programme and demonstrated this by making resources available to us (time and money!) as well as going out on the plant to undertake observations. We were very committed to the programme and delighted to see how frontline employees, contractors and some of the management team embraced it.

Moreover, there are lots of examples of communications, many of those humorous, which we used throughout the programme. We used comedians from TV shows in our posters and designed funny messages; it almost resembled more of a marketing campaign than a safety programme. The posters were changed every couple of weeks as we wanted to keep the messages alive and have the posters serve as conversation starters. Do posters make your workplace safer? No, is often the answer. Do posters keep the safety conversation alive if they're humorous, topical and changed regularly? Yes, at least in my experience. We went so far as to get special toilet paper printed stating "you wouldn't start this job without the right paperwork" at one stage when our focus was on permits. As well as a nice graphic representation of that message, we also placed a poster on the back of the toilet door with an appropriately thick brown crayon font. We unapologetically barraged people with reminders of safety, even when they went to the loo!

So, where did this all lead to? Was it all rosy in the garden? Was it plain sailing off into the sunset with our programme? It certainly led to some positives, but it also revealed some unexpected challenges, some of which led to obstacles that really surprised us as a committee, but more on that later. First, let's look at the positives. We had our pilot review after 12 months, and the approach seemed to have a positive influence. Although only half of the site was on the pilot, the overall site stats measured by lagging indicators took a plunge in a desirable direction. Maybe this was a bit of a butterfly effect; the fact that our communications were site-wide meant that all plasma screens on site showed our weekly updates. What would have previously been perceived as bad news was now being displayed across the site as something to be celebrated. People were thanked and asked to keep highlighting any areas where they saw the potential for someone to get hurt. There were also some people who, while mainly based in the pilot area of the site, also moved in their roles to the other plants on site. Maybe a heightened awareness or willingness to raise matters emerged from that, we can only surmise.

For over ten years, the site had plateaued regarding Lost Time Injuries (LTI). After one year in our programme, this number dropped by over 50%. When you looked at the chart, the decline was pretty dramatic. It is worth noting that the near-miss reporting system stayed the same and was complemented by the observations undertaken and submitted. While our goal was for employees to "go home safely", we weren't focused on the numbers. It came as a shock to us when the plant manager told us it was the first time that the plant had been "injury free" since its operations began. I'm well aware of the fallacies of the quest for zero and the cobra effect and was delighted to think that via showcasing safety in a "warts and all" way (the good, the bad and the ugly aspects of it on our site showcased for all to see), we achieved fewer injuries by wanting to know more and encouraging the sharing of that information.

Going beyond numbers, there also appeared to be a shift in the perception of safety on site. Those that raised red flags with the union on the launch of the programme were now requesting observations to be undertaken on tasks they were carrying out. When these originally cynical folk bought into the programme and saw value in it, that represented, for the committee at least, a real shift in how our programme, and safety in general, was perceived and felt. Furthermore, there was one work shift that had been heavily impacted by a historical incident, whereby one operator challenged another on the grounds of safety and it became an HR issue. Consequently, on that shift, nobody challenged anyone as a result of this precedence. Twelve months into our programme, feedback from an observation was overheard being given between two members on that shift, some of which highlighted some gaps as well as positives; this conveyed the success of this programme in a small, yet monumental way. When we can facilitate and make it safe for conversations like this to happen, contributing to psychological safety, we're doing something good. Trying to reflect that on a Key Performance Indicator (KPI) though can be a problem; that is something that managers struggle with, as in "how do I measure that?" Obliquity is key but measuring can prove troublesome in the short term.

Externally, we also received some encouraging feedback. The organisational psychologist that helped us had the following to say: "I have being involved in a lot of change programmes in many companies over the last 25 years – for uniqueness, innovation and participation this programme is the most successful I have experienced – well done to you all". A delegate of a conference I was asked to present on had the following to say: "Can you please ask that wonderful presenter Marian Kiely for her presentation on BBS. She was inspirational, and I loved the fact that the project came from the grassroots of the organisation as she so aptly put it. Her presentation was informative, witty and unique in that it focused on the positive rather than the negative blame culture that we see so often with Safety. Please pass on my praise to her and I hope she won't mind us getting in touch in the future".

Ranking risk was another area where there was a bit of discussion and debate around. We liked to look at our risk matrix as being 3D rather than 2D, like a Rubik's cube, so to speak. Traditionally, when work orders were being prioritised, the risk rating was based on severity and frequency. We added a third dimension, that of "frontline emotional litmus test" (FELT). We based this third dimension on how vocal and frustrated frontline were with any given aspect of a task. If they were very engaged and vocal, and it was widely felt, this gave more weight to that item. To give an example, when taking a sample from a dryer, sometimes an operator would find no breathing air hose to attach their air hood to carry out the task safely. They would have to go and try and find a hose elsewhere and this could take up to an hour, depending on the availability of hoses. While the risk from a management perspective was minor, as it was only a matter of getting a hose elsewhere, the frustration experienced by operators was very high. They would have to spend time finding a hose in another room; if they found one they would have to ensure it was clean or clean it before bringing it to the required location, and at the end of the shift, they would have to account as to why their process wasn't where it was expected to be if there were any delays.

Prioritising what were perceived as minor items like these did a lot to build trust in the programme as frontline employees felt valued and respected relating to items raised and actions taken. As the plant manager mentioned at the time: "This programme has had a very positive effect on safety and safety engagement in the pilot plant, greater than any other single safety initiative onsite … I've noticed a significant change in the pilot plant safety meetings, there is more partnership, and the programme has cleared out many of the small issues that used to dominate the meetings and cause so much frustration and conflict".

It is also worth noting that many procedures were highlighted as being incorrect as the programme progressed. This resulted in over 40 procedures being reviewed by the training department in the first year, with the input of frontline employees, the quality department and management. This wasn't a methodological approach to review procedures, but a natural trickling out via observations and highlighting a conflict in how the procedure outlined how to do a task and the way in which it was done. This resulted in reducing some previously unknown risks present in relation to ergonomics and chemical exposure, amongst others. Today, the term organisational drift is used to describe this, highlighting the difference between "work as imagined" (WAI) and "work as done" (WAD), in the words of Professor Erik Hollnagel. Through an organic way which sought the input of many stakeholders in any given process, the gap between WAI and WAD was being narrowed.

Now, to the more negative aspects of how the programme was received. The committee expected that senior leadership and middle management would be supportive of a new grassroots safety programme and actively do what they could to promote and support it. We thought resistance would come from frontline employees, from getting them to buy in to observing someone or allowing someone to observe them. We were proved wrong; in fact, much resistance came from middle management and senior leadership. To be fair, some managers and leaders were advocates and supported the programme, but others actively went about detracting from the programme and openly knocking it. The steering committee was not expecting this; as mentioned earlier, we were very naive in some ways.

Although we had representation from the Health and Safety (H&S) department in our team, we learned as we went that the H&S department were not fans of ours. The beat on the street was that they felt that our communications, approach and results made them look incompetent; hence, they actively went about putting us down to others. At the time, the steering committee took great offence to this as we saw ourselves doing a lot to bring about positive change in safety and yet we received no recognition or encouragement for our efforts from the H&S folk on site. Now, going back to Johnny Cash's "I can see clearly now" lyrics, I can understand it. But back then we didn't get it at all. Another area that proved to be prickly was dealing with the engineering department who processed work orders that arose from observations. Due to the ramping resistance from engineering supervision, which was often vocalised quite abrasively, it proved very difficult to get work orders relating to observations completed.

Now, with the benefit of hindsight, it is well understood why these groups did not relish our arrival on the safety scene. The H&S department had previously raised some of the items which were now rising again through observations and had requested some resources to address them at the time. They were not given the support the steering committee received. One can well imagine their frustration looking on as we posted weekly updates to the plasma screens across the site on solutions and improvements we were implementing. Looking at the world through our lenses at the time did not reveal this; we were too caught up in what we were doing and they were too offended and enraged to discuss them with us.

Concerning the engineering side of things, everyday items needed to get done as well as items raised through our observations. However, every department in an organisation is given a certain amount of resources and must manage their work in line with the budgetary and time limitations accordingly. Later, it became clear that not enough extra resources were given to the engineering department to complete the snowball of work emerging from our observations, yet the expectation was that it would all get done. When the steering committee got frustrated due to inaction, and the engineering supervision, in turn, got frustrated with that frustration, we were going nowhere, and we were going there fast. At times it was a case of "the squeaky wheel gets the grease", and not making enough noise led to inaction. We were all seeing things from our own perspectives, us seeing a six and them seeing a nine (or vice versa), perceiving the other as wrong, but not willing to look at it from each other's perspective. Soon, it became a battle of wills, and at times a shouting match, which sometimes ended in tears. How often do we see this in our organisations? How often do politics and egos come in the way of understanding and moving forward? How often do we see "the tail wagging the dog" whereby weak leaders allow direct reports to negatively influence the direction of their department without intervening?

Looking back now, I can see that our biases shape our behaviour. We jump to conclusions based on fragmented pieces of information which we categorise based on patterns we come to recognise through our lived experience. Having come from the grassroots via an operator role, I was tethered to that space by constraints not visible to me at the time. The same could be said for a supervisor who is tethered to their space at management level by constraints not visible to them, yet these invisible and intangible constraints shaped and influenced how we engaged and interacted with each other. The committee was so blinded by our allegiance to the frontline group that we failed to see things from the management's perspective. Our biases told us that our challenge was in convincing frontline employees of the value of our programme, we failed to see the glaring omission of convincing management of the value of it. Seeing the world as always being an ordered space was something I was also guilty of. The leadership and management teams were given a talk by the site leader about the value of our programme, so we didn't have to worry about that side of things, right? As we later found out, that was a costly assumption. Although our communications were very good at delivering information, we might not have been as good as we'd have liked to think on the listening side of things. That is definitely a takeaway and learning.

Today, I need to ask what I would do differently. It's funny when we're in the middle of the storm we can't see what's right in front of us, and we often think that no other organisation would have an issue like this, accept people who behave like this or no other organisation could be that political. The truth is, I have come across this conflict time and time again regardless of being in the public sector or private sector, multinational or small and medium businesses. The conflict that comes from the breakdown of interpersonal communications, where people are more interested in talking than listening and more interested in being right than learning, is widespread. When it comes to H&S, there is a growing recognition of the need for softer skills. A lot of emphasis in the past was on technical skills; while this is important, our changing world of work requires a different skill set which incorporates coaching capabilities, emotional intelligence, awareness of the bigger picture and complexities at play, and self-compassion.

Since my time in that role, I have immersed myself in anthro-complexity theory and applying it to safety. I have learned that when we are dealing with complex issues, to drive on with a solution-oriented focus often does not serve us well. There may well emerge "unknown unknowns", some with undesirable outcomes. It is imperative to probe first by bringing a cross-functional group together and making sense of a situation and deciding with the perspectives of all involved taken on board. A grassroots initiative is all well and good, but it can't fly without the support of senior leadership and middle management. To involve these people from the outset and take on board their concerns and ideas would be well worthwhile. To whet their appetite for any proposed change, I would consider it essential to involve them at an early stage and have them feel a valued and respected part of the team.

As a facilitator to the programme, I was very dedicated and persevered to the best of my ability to fulfil my role in ensuring everyone went home safely. Being honest, again in hindsight, I was ridiculously over-committed. This over-commitment to the programme led to long working days, which turned into long weeks and months, regularly working 60–70-hour weeks. We have all worked on projects for long hours for short stints; sometimes it is needed, but when it becomes the norm, it is far from ideal. I would often leave home at 6/6:30 am and not get home from work again until 10/10:30 pm; long working days became the norm. I would return home still buzzing from my day. The only thing that would help me switch off was having dinner in front of the TV with wine for company. Dinner was not some beautifully crafted meal cooked from scratch as I did not have time for that. It was more often than not a pizza or a takeaway; I would chill for a couple of hours, get some sleep and the cycle would start again the following day.

I would like to think that I am tough enough that I can withstand folk constantly nit-picking and looking for fault; but, when this happens day in day out, you wear thin. The thick outer shell becomes weak and vulnerable, and cracks begin to appear over time. Having been in the role for almost two years, I found myself on a train coming back from Dublin one weekend alone, after spending time away with friends. The previous week at work was very stressful for a number of reasons, one of which was receiving a highly charged e-mail, with many of the senior leadership cc'd. Looking back now it shouldn't have upset me as much as it did, but at the time it was the straw that broke the camels back. I couldn't continue as I was, I

felt alone and hopeless, the tank had run dry and I had no more to give ...
Out of nowhere, the tears came. I could not stop. I did not know what was
wrong and I was glad to sit in a quiet carriage so that I could cry without an
audience. I arrived home and joined some friends for dinner that night. The
tears came again; I just could not stop. I did not know what was wrong. My
friend suggested I go to a doctor the following morning and take a week off
work. I made an appointment the following morning and racked with guilt
rang into work sick. I could hardly articulate to the doctor my ailment; such
was my upset. He suggested I take two weeks off, which I did. I ended up
being diagnosed with burnout and work-related stress and I was out of work
for almost six months. I have since learned that burnout is a new word for
a mental breakdown. Wow! The one that was out there protecting everyone
else left herself get really hurt!

Looking back at that time, work consisted of a heavy workload and
no-one to share it with, little influence over what got done, a little but not
enough support from peers and leadership as our interactions were brief
and few, being the recipient of hostility and aggression at meetings repeat-
edly and all this paired with "all work and no play" meant that it wasn't a
matter of if, but when, I'd no longer be able to continue. Being out of work
sick resulted in feelings of guilt and added stress as I thought the organisation would fall without me, but as it turned out, they survived just fine.
Having been through burnout, I would not wish it on anyone, but it taught
me a lot and led me to question everything in life. Being truthful, I am
lucky to be here today as a result of going through it. I appreciate the learnings from this experience and they stand to me in everything I do since. I
questioned why I was living where I was living, why I was spending much
of my time in my chosen place of work, why I invested time in friendships,
some of which did not deliver a good return on investment of my time and
so on. Much soul searching and reflection went into finding answers to
these questions and this guided action around career, relationships, education and so on.

When being asked for a story for success and one of failure, I combined
both into one. I worked on a very successful programme which in hindsight
had many aspects of Safety Differently, and it without question prevented
people from getting hurt. That was a great success. The failure? That lay in
not pre-empting the resistance from various functions within the organisation and not recognising it in the early stages and doing something about
it. There was definitely learning in that. This lesson taught me that with the
best of the intentions in the world, if a grassroots initiative is not met with

support from senior leadership, chances are it won't end well. Politics and organisational psychology are not to be underestimated.

Looking ahead, what has this taught me? Priming the organisation and setting the scene I believe are very important initial steps, along with outlining to each business unit the expectation of them that is required to make the programme a success; joining the dots as to what success will look like and their part in that. Storyboarding and showing "this is how it is, but this is how it could be", can help some seeds take a shoot in relation to the programme vision. Gaining multiple perspectives from the outset will help with sense-making, and by taking these perspectives on board as the programme sets out, it will also cultivate ownership in the programme. Coaching skills would be very beneficial for those trying to introduce a new programme, paired with an awareness of organisational psychology and anthro-complexity.

Busyness has now almost become a badge of honour in some organisations and many people are struggling to cope with the various work and home life demands. When facilitating a sense-making workshop some time ago, one of the participants gave feedback that to take time out to hear each other's perspectives and deepen their knowledge about health and safety was a luxury they were rarely afforded. In the story I have presented here, everyone was doing their best, motivated by whatever it was that their success was measured on. Imagine if we had taken time out at the start of this journey and at various stages in between; where it might have led us, we can only guess ... Is taking time out every now and again something we can't afford to do? I should think not ...

If I were to give advice to a younger me setting out on this programme, it would include the following:

- Attain a coaching skill set; it will help set you up for success as you go.
- Listen more and talk less, eco as opposed to ego. If you want to influence the ecology of your organisation, you need to understand how people see and perceive it.
- Secure autonomy in relation to making decisions around action required, having to go cap in hand continuously is a recipe for disaster.
- Set clear expectations from the outset, let leaders and managers know that their decisions may well fall under observation if it is felt they introduce unacceptable levels of risk to the organisation. If we really want to bring about a true just culture, this cannot just be peer-to-peer discussions and observations.

- Be mindful of those whose video is not aligned with their audio, see through the BS and be mindful of it as you go, know who you're dealing with.
- Zoom in and out, try and see the bigger picture as well as the detail of the task in hand. Being too focused and having the head down continuously can result in losing sight of the bigger picture and prove costly.
- Don't always be afraid of silence; at times, dis-ease can sometimes lead to valuable input from others.
- Mindfulness and a growth mindset are invaluable as a means of harvesting and protecting your energy. Be mindful of those who will zap you of yours, invest in relationships that will amplify rather than dampen your energy.
- We need to harvest psychological safety and welcome variety and challenge – premature convergence/surrounding ourselves with people who agree with us all the time is not desirable if we wish to bring about meaningful and lasting change.
- Don't be too hard on yourself; you are doing your best. It is not your role to "keep the peace" in every instance; mind yourself and practice self-compassion.

Chapter 12

Learning from Incidents: Mind the Whole Set of Dimensions

Mark Sujan

Contents

Healthcare is a risky business. In recent years, several high-profile incidents have hit the headline news and mainstream media. For example, the public inquiry into the appalling care delivered to many patients at Mid-Staffordshire NHS Foundation Trust was followed closely by the media. Similarly, the threat arising from hospital-acquired infections is well documented, and patients and the public understand the importance of hand hygiene in hospitals. However, what many people are unaware of is that beyond these well-publicised incidents, which are the tip of the iceberg, thousands of patients are harmed every year. Examples include patient falls in hospitals, medication errors, retention of surgical instruments in the body, unrecognised deterioration, and diagnostic errors. A widely quoted figure suggests that as many as one in ten hospital patients will suffer an adverse event during their treatment, with about half of these thought to be entirely preventable, hence causing unnecessary harm and distress to patients, their families, and the healthcare professionals involved in their care.[1]

Hospitals and other healthcare providers have a duty by law to investigate serious adverse events and many organisations operate incident reporting systems. Most healthcare providers choose Root Cause Analysis (RCA) to generate learning from incidents. RCA is the accepted approach to deliver this, but other methods could be used if an organisation wanted to make an argument for it. RCA aims to clarify what happened, identify contributory factors, and develop recommendations for improvement to practice.

Literature documents quite clearly that healthcare organisations are struggling to generate learnings from incidents that are relevant to clinical practice and lead to sustainable change.[2,3,4] These studies document barriers to effective incident reporting in healthcare, such as fear of blame and repercussions, poor usability of incident reporting systems, perceptions among doctors that incident reporting is a nursing process, lack of feedback to staff who report incidents, and lack of visible improvements to the local work environment as a result of reported incidents. I have similar experiences in my own work with hospitals. I work with hospitals as an external consultant in patient safety and quality improvement, often funded via a third party such as a charity or an NHS body within the Department of Health.

When I speak with clinicians about learning from incidents, their responses are frequently very similar and sobering. Many clinicians say that they do not understand how incident reporting works in their organisations, and those who contributed incident reports in the past feel that these reports somehow disappeared into a black hole with little feedback or change happening as a result. Reflecting on my experiences of working with healthcare organisations over the past 15 years to improve patient safety, I felt that key challenges for learning from incidents in healthcare would be to address the disengagement of clinical staff and avoid negative connotations of human error and blame.

To this end, I started working with hospitals on an approach to learning from experience that could be owned and managed by frontline staff and departments, and that could produce actionable learning of direct relevance to people's clinical practice. The experiences I draw upon in this chapter are from a six-year project called Safer Clinical Systems (SCS). The SCS project was funded by the Health Foundation, a UK charity. The project included 14 participating hospitals. Four hospitals formed the pilot, eight hospitals participated during the second phase, and another two hospitals were recruited as part of the validation phase of the learning from experience work described in this chapter.

The overall aim of SCS was to co-design and test a systems-based approach to patient safety improvement, largely inspired by safety management practices from safety-critical industries. The approach included the systematic identification and assessment of risks and the development of a safety argument documented in a safety case.[5] More specifically, in my stories below, I reflect on the early successes of developing and implementing the approach in a hospital pharmacy, and the more mixed findings when the approach was rolled out more widely. I believe this provides a useful illustration of the importance of understanding what works, for whom, and under what circumstances, because the successful adoption of any safety improvement and safety management approach depends as much on the path itself as it does on the details of its implementation.

Learning from Hassles

Having a great team is the foundation of success. In the case of our learning from experience pilot, having a dedicated clinical team with the right level of authority and senior management support proved to be an essential ingredient for developing and implementing this new approach. I started working with the hospital pharmacy team at a district general hospital in 2008. The core pharmacy project team included the director of the pharmacy, the lead technician, and several dedicated pharmacy staff, supported by the medical director and the quality improvement lead. The pharmacy team wanted to improve their practice and enhance patient safety. In early discussions, they suggested that they had done a review of all reported incidents, but there were, in fact, few incidents that had been reported. Also, the level of detail provided in the incident reports was insufficient to allow the team to generate any meaningful learning.

I pitched them the idea of implementing an approach to learning from experience based on eliciting narratives about hassles from frontline staff. This type of approach became known as "hassle reporting" within the pharmacy. Learning from hassle is a complement to the mandatory investigation of serious adverse events. On the one hand, when something goes wrong, and a patient is harmed, there are issues around responsibility, accountability, and blame that need to be carefully navigated. On the other hand, people experience hassles daily, and they are usually more than happy to share their hassles with people who are willing to listen. Importantly, though, analysis of hassle narratives provides useful insights into latent organisational factors that can subsequently contribute to adverse events. Examples of hassles are issues such as staff shortages, poor usability of equipment, broken printers, and inadequate procedures. Box 12.1 below provides examples of short descriptions taken from submitted hassle narratives.

BOX 12.1   Short Examples of Hassle Narratives and Underlying Latent Organisational Factors

Example 1 – Staffing levels (absences, staff shortage, inappropriate skill mix): "It was just one of those days where I felt that I wasn't getting anywhere, we were short-staffed due to sickness and annual leave, the phones never stopped ringing, and sometimes I think I am the only person who can hear them ringing".

Example 2 – Work environment (insufficient space, untidy, interruptions): "Folders left out, and bits of labels all over the workbench made work difficult as no space. Leaflets, meds and bits of labels etc. do get left on the bench, but it happens when we are busy, and everyone has different ways of working, it can be very untidy sometimes".

The pharmacy project team were very keen on developing and testing this idea for a novel approach to learning from experience to improve patient safety. The lead technician acted as the hands-on local project lead. She had excellent working relationships with pharmacy staff as well as with staff from other departments who frequently interact with the pharmacy. I had regular meetings with her and provided theoretical and practical support for safety and human factors. The pharmacy director and the wider project team joined us for monthly project meetings. During the project meetings, updates were presented, and any barriers to progress were discussed. Having the pharmacy director as well as the medical director and the quality improvement lead on board helped the project team to raise issues that required senior management sign-off and support.

The pilot ran for 18 months. During this period, many hassle narratives were submitted by staff. The pharmacy director authorised the introduction of a "safety time", which is time individual members of staff can take out of their regular working hours to contribute to the hassle reporting project. This proved to be very important because employees had suggested that they did not routinely have time available to contribute to patient safety improvement initiatives and that frequently they had to use their lunch break or stay on after their shift for this purpose. The safety time was a 30-minute time slot set aside once a fortnight, which could be rostered on an individual basis (i.e. not everyone would take it at the same time). Of course, word demands do not lessen and staff numbers do not miraculously increase. However, the introduction of the safety time was a visible sign of commitment from senior management that this activity was valued. Colleagues, as well as management staff, would cover someone's shift during the period where they were taking their safety time.

The second significant organisational change was the introduction of ownership and authority for implementing improvement actions that were derived from the analysis of the hassle narratives. Previously, such initiatives usually resided with the quality improvement or risk management departments. In an attempt to bring the learning closer to the actual clinical frontline, staff were encouraged to take on responsibility for driving forward improvements that related to their practice, and senior management provided support accordingly. Such improvements did not have to be only larger strategic projects, but shorter "easy wins" were also encouraged. Each month, feedback to all members of staff was provided in staff meetings, and in this way, demonstrable progress was made.

Many safety improvements were introduced and tested throughout the project. An example of a longer-term strategic improvement was the resolution of IT problems. A review of all IT within the pharmacy was undertaken and the communication channels with the IT department were reviewed. The IT department fully engaged in the process and provided feedback and clarification on the procedures and processes for reporting IT-related issues. This improvement project required communication across departmental boundaries and was, therefore, by default of higher complexity and required a longer time frame. An example of a faster improvement project (one of the "easy wins") was in response to the reported problems with the cluttered and untidy work environment. A dedicated space was created in the dispensary where the pharmacist could undertake final checking of all drugs that were dispensed. This is an important safety check and getting it right is crucial. The problem was identified, analysed, and a consensus was developed within a month, and the improvement was implemented using Plan-Do-Study-Act (PDSA) cycles the following month.

The "plan" was to create a space within the dispensary for double-checking by moving the card payment system and the medications waste bin to a different area, reviewing and tidying the shelving space, and setting up a dedicated space for a laptop in this area. The "do" phase consisted of trialling these arrangements quickly the following month. The "study" phase of the cycle consisted in recording the number of times the pharmacist needed to use a dispensing station rather than the clinical check station, and eliciting staff feedback (e.g. about potential negative consequences). Following positive feedback, the "act" part of the cycle consisted of making these changes permanent. Usually, several PDSA cycles are required to implement an improvement. However, this problem had a reasonably straightforward solution, and staff were happy that the solution was working after the improvement cycle was completed.

What made this approach work? As part of a research project, I undertook detailed interviews with staff about their experiences with the approach. The feedback was very positive. Staff suggested that the hassle narratives were a very useful way to contribute to the system for organisational learning through their experiences. The hassle narratives also provided a means of reflection on working practices, both one's own as well as departmental working practices. Staff were very enthusiastic about the visible improvements that took place in their working practices, which was put down largely to the "easy wins" improvements, but which then acted as motivation to continue to engage with the process.

And as I mentioned in the introductory paragraph, having available the right team which was able to engage with staff and had sufficient seniority to provide support to make changes happen was the critical success factor. The active support of the director of the pharmacy was vital. It ensured that staff were able to contribute to the project, that they could set aside time, and that authority was given to make changes happen. The presence of the director of the pharmacy in team meetings and during staff updates sent a strong signal of departmental and organisational commitment to the project and improvement in patient safety more generally.

The director of the pharmacy also communicated the project at the hospital board level and other departments, an initiative which created awareness, interest, and a willingness to collaborate across departmental boundaries. This was important on several levels. First, many issues require solutions that stretch across the boundaries of a single department. Second, the hospital board needs to demonstrate a commitment to safety and quality as part of regulatory requirements and having such initiatives can help them show this commitment to the regulator. Equally important was the contribution of the lead technician who acted as the project coordinator. Initially, the time of the lead technician on the project was funded externally through the research grant, but subsequently, the pharmacy department took over the funding for this. Again, this would not have happened without support from the director of the pharmacy and the hospital board.

A Disease That You Just Cannot Seem to Shake Off

A dysfunctional culture and lack of trust can easily kill off the best-intentioned improvements. People need to want change, and they need to trust their management that they are ready for change too. When promises are not followed up with actions, people will say that they have heard it all before and start disengaging from improvement and change processes. As one nurse put it to me in a conversation about her work environment, there was a feeling of resignation among her colleagues that felt like a disease that you just cannot seem to shake off.

Armed with the success of the pilot study described in my story above, the funders of the work felt that the approach was ready for roll-out, and we certainly felt very encouraged. The learning from hassle approach was adopted in ten hospitals in 2011, two of which were selected for in-depth research about the performance and utility of the approach in diverse environments. One of the study environments was quite close in set-up to the pilot hospital. The department was radiology and diagnostics, and the team included the head of the department who was also the medical director, a radiographer as project lead, and several enthusiastic members of staff. The second study site was a hectic surgical emergency admissions unit, hence very different from a pharmacy or radiology department. The team set-up was also very different, with the team leader being an enthusiastic surgical trainee, supported by a research nurse. The project was approved by the head of surgery, who, however, was not part of the project team.

Differently from the pilot project, I provided less hands-on support to the hospitals. I held a training day for all hospitals, and then I visited the two study hospitals once a month throughout the 18-month roll-out project. During these meetings, I acted as a critical friend and advisor, but project teams were free to implement and use the approach in a way that seemed most applicable and fruitful in their environment. The reason for taking this approach was twofold. On the one hand, we recognised that learning from hassle would need to be tailored to the specific setting. On the other hand, we needed to test whether such an approach can function without close supervision by an external expert, meaning whether non-specialists could own and run the approach.

Implementation in the radiology and diagnostics department followed the path of the pilot study fairly closely and similar success was obtained. They adopted the safety time concept and provided monthly updates to staff about improvements. The nature of the improvements was slightly different. In the pilot study, there was a lot of emphasis on the "quick win" types of interventions to sustain continuous visible progress and foster staff participation. In the radiology and diagnostics department, the focus was predominantly on a few selected strategic improvements. This was because the project team felt that these were of particular importance and they did not have sufficient resources to address other issues. They also had a less active contribution to improvements than the pilot study site, where the "quick wins" were implemented predominantly by staff who reported hassle rather than the core project team.

In conversation with the project team, they suggested that the difference above might be because, in radiology, people are working more independently and more on their own compared with, for example, a pharmacy environment. Hence, the project team suggested they found it more challenging to involve staff actively. However, there were some improvements led by frontline staff. So, my impression is that it is probably more a matter of departmental culture that determines the extent to which staff are encouraged to contribute actively rather than necessarily the type of work a department does.

Despite these differences in implementation and focus, the "learning from hassle" approach generated useful, actionable learning, which was then implemented in patient safety improvement interventions. The main strategic improvement activity was around addressing the communication with theatres requesting radiographers to support ongoing surgery with mobile imaging equipment. This communication is time-critical and was felt to be inadequate. Requests for radiographers often come in at short notice and are usually not coordinated as they can originate from different specialities. As a result, the central radiology department might be left without appropriate cover and appropriate supervision arrangements for junior members of staff. There might also be delays in performing the imaging in the theatre because there is only a limited number of mobile machines available and this can cause delays in surgery and affect patient outcomes.

Communication across departmental boundaries in a hospital is never an easy matter to address due to differing priorities and an unclear allocation of responsibility. For example, in this case, the main priority of the theatre manager was to schedule elective and emergency surgery as efficiently as possible to minimise risk and delays to patients. For the radiology department, on the other hand, there were other priorities to consider, such as providing continuous service without delays to the emergency department as well as to theatres. With the evidence generated from the analysis, the radiology team felt well prepared to initiate a dialogue with the theatre manager to raise awareness of this issue. Subsequently, an electronic booking diary and a standard operating procedure for booking the mobile imaging equipment were developed. This was supported by an inter-departmental working group, which was established specifically for this purpose.

The implementation of the project in the emergency surgical admissions unit was very different. The project team lost important senior members early on due to other commitments. The surgical trainee, who was leading the project, left the hospital halfway through the project when his rotation came to an end. He continued to be involved through dialogue with the research nurse. Still, in essence, this meant that for a significant amount of time, the project team did not include anybody actually working in the department. The department was also undergoing structural and organisational change. The most significant change was the move to a new physical location and the merger with another department. This created uncertainty among staff over several months and resulted in significantly changed working conditions and staff composition. Subsequently, the two departments were separated again following a regulatory intervention. For a large part of the implementation period, this resulted in highly unstable and uncertain conditions in the department, which also led to breaks in the implementation of the "learning from hassle" approach.

These organisational difficulties notwithstanding, the approach generated some patient safety improvement interventions that were implemented successfully. These related more to the "quick wins" interventions, because the project team did not feel they had sufficient support nor authority to address more strategic issues. The patient safety improvement interventions were mainly around the work environment and fixing missing or broken equipment. However, the approach was not sustained due to the lack of buy-in and absence of ownership by staff, and the lack of senior management support.

Drawing these experiences to a close, a key lesson from the implementation of a novel approach to learning from experience in healthcare is that several essential implementation factors greatly influence the extent to which the approach can contribute successfully to organisational learning and patient safety improvement. These factors are around staff engagement, senior management support, implementation team composition, and organisational readiness for change.

Active staff engagement can lead to more proactive learning, a sense of empowerment, and contribute towards positive staff morale. Key strategies for engaging staff include adequate communication and feedback, collaborative solution development, and harnessing of professional incentives. Obstacles in engaging staff are loss of continuity due to frequent staff changes, low staff morale and negativity, staffing levels and workload, slow pace of change, and complicated relationships across departmental boundaries.

Senior managers can enable change. They can also engage at a managerial level with stakeholders across departmental boundaries. Critical strategies for securing top management support are early involvement of senior managers and data-driven communication backed by systematically gathered evidence. Obstacles in obtaining senior management support include the lack of priority for proactive learning and improvement, the personality of individuals and their interests, and the high turnover of senior management staff, which makes establishing working relationships difficult.

Adequate team composition facilitates staff engagement and contributes to securing senior management support. In this way, it also contributes to the successful implementation of improvements. Strategies for building an appropriate team include the assembly of a large, multi-disciplinary team with different strengths, the inclusion of ward champions to ensure a continued presence in the work environment, and the early involvement of senior managers in the team. Obstacles include inadequate staffing levels, frequent staff changes resulting in a lack of continuity of team membership, a limiting part-time role that reduces the presence in the work environment, and a lack of senior management involvement due to other priorities.

Proactive learning and service improvement should become a core function and priority of the organisation. To influence senior decision-makers, it is crucial to present quantitative evidence and communicate what the benefits to the patient and the organisation would be. Obstacles include productivity pressures, low staff morale and negativity, and a predominantly reactive approach that keeps the organisation in a continuous mode of having to manage crises.

It is, therefore, important to consider not only the technical aspects of safety management approaches, such as the approach to learning from experience, but also the organisational and implementation dimensions. Evaluation of safety management approaches requires careful investigation of the mechanisms that bring about improvements and the context. Evaluation needs to shed light on what works, for whom, why, and under what circumstances.

Notes

1. Vincent, C., Neale, G., Woloshynowych, M. Adverse events in British hospitals: preliminary retrospective record review. *BMJ (Clinical Research Ed)* 2001;322(7285):517–9.
2. Kellogg, K.M., Hettinger Z., Shah M., et al. Our current approach to root cause analysis: is it contributing to our failure to improve patient safety? *BMJ Quality & Safety* 2017;26(5):381–7. doi: 10.1136/bmjqs-2016-005991
3. Peerally, M.F., Carr, S., Waring, J., et al. The problem with root cause analysis. *BMJ Quality & Safety* 2016. doi: 10.1136/bmjqs-2016-005511
4. Sujan, M. An organisation without a memory: a qualitative study of hospital staff perceptions on reporting and organisational learning for patient safety. *Reliability Engineering & System Safety* 2015;144:45–52. doi: http://dx.doi.org/10.1016/j.ress.2015.07.011
5. Spurgeon, P., Flanagan, H., Cooke, M., et al. Creating safer health systems: lessons from other sectors and an account of an application in the safer clinical systems programme. *Health Services Management Research* 2017;30(2):85–93.

Chapter 13

Necessary Incompliance and Safety-Threatening Collegiality

Nektarios Karanikas

Contents

My stories below refer to the period of my service as a chief engineer in an intercept aircraft squadron. I was responsible and accountable for the performance and safety of about 100 engineers and technicians (officers and non-commissioned officers), and the availability and airworthiness of military aircraft and their supporting tools and equipment. The operational profile of such aviation military formations dictates the maintenance of the maximum possible availability of airworthy aircraft for training and interception missions.

When Trust in the System Is Missing

It was a typical morning while arriving at work when I noticed that the back cabin of a support vehicle was bended and malformed. The specific vehicle was typically operated by technicians belonging to one of the three technical branches I was managing; the particular branch was responsible for first-level aircraft maintenance (e.g., troubleshooting, replacement of parts, ground tests) and they were using the vehicle to commute to aircraft shelters and transfer parts and equipment within the squadron or other places on the military base. However, when necessary, the vehicle could be used by other technical branches (e.g., in case of malfunction of their own transport means).

When I saw the problem, I thought that something serious must have happened, which I was not aware of, although, theoretically, my staff should always report to me anything "abnormal" as soon as possible. This hypothesis of mine, and all managers typically, was already disproved. I started asking around for information, but everyone declared themselves unknowledgeable of what, where, and when any incident had occurred, and which persons were involved. Although, on the positive side, nobody had been injured (i.e., I would have been informed by the medical services and the squadron commander), it was necessary to learn more about the event. I wanted to avoid similar occurrences in the future, and, rather equally important, inform the squadron commander, whom I fully trusted for his leadership and management skills.

However, every office and workshop staff I asked declared themselves ignorant. I felt uncomfortable: was it a matter of trust in me or the broader system? My second hypothesis about trust of my staff in me was not yet evidently tested, but the indications were against its standing. Nonetheless, what were the boundaries between the system and me really? I was the "system" for my subordinates. I had the option to report the event upwards in the hierarchy, to the operating base level. However, my experience from the treatment of such events in the past was unpleasant. The bad news would spread quickly, the safety and security personnel would make a visit on the squadron to find "facts" and identify the careless people involved, and the base commander would criticize us for not respecting our technical assets, inadequate supervision, and mismanagement. This is what had happened plenty of times in the past with the same actors; the context never played a major role. Hence, this third hypothesis was well-substantiated; I decided to become "incompliant", at least until my "internal" investigation concluded.

My very first decision was to secure the damaged vehicle and impose strict and full control over the movement of the rest of the vehicles. The officers of the Maintenance Control Office were to keep the keys of all vehicles, deciding who would operate the vehicles and recording locations, times, reasons for movement, etc. This decision discomforted my staff because it caused delays in the completion of their tasks, but they understood that my trust in them had lowered. I did not "think" reciprocally; it was more a naturalistic reaction rather than a rational one. I thought that I had to try taking control of the situation until I had figured out what had happened with the specific vehicle. It seemed like the acute reactions in the past when we had an aircraft crash, after which the whole fleet of the respective aircraft type was grounded until the investigators had collected the first data and developed a hypothesis of what might have happened. If the major problem had been a technical failure, then we should check all aircraft before releasing them for flights. I was rather driven by this practice, and I applied the same to my case.

Over the next few days, I continued with the rest of my normal tasks, I remained calm, and I was hoping that someone would knock on the door of my office to confess or report about the particular event. I had clearly promised that nobody would suffer from negative implications, but nobody appeared. My next decision after discussing with the squadron commander was to lift privileges that we had honoured to the employees over and above standard procedures and rules. For example, the duty-rest period scheme in place required technicians to not report for work earlier than 12 hours after they had worked overtime and to perform no maintenance task earlier than that. Instead of complying strictly with this policy, our squadron strategy was to award a day off (extra day of paid leave) to everyone who had to stay longer at work for any reason (e.g., bring back in service a defected aircraft, service night flights) regardless of the number of overworked hours. This motivated our staff who were coming to work rested after 1.5–2 days and did not have to pay their transportation costs to commute to work the next day at times of the day when there were no alternative options (e.g., riding the free bus in the morning). At the same time, this policy was on the safe side (i.e., we did not ask staff to get back to work earlier than the rest period) and served our mission well to ensure the maximum possible aircraft availability; it was a win–win situation. We had a few similar policies in place in the name of good leadership, management, and relationships.

However, it is important to note that some of these privileges were somewhat "incompliant" with established rules, meaning that the commander and I stretched the provisions of procedures and demonstrated trust in our staff that they would not behave in a way that would raise top-down questions about these differences. Lifting these extra "employee privileges" was once more an expression of lowering our trust in them. We knew that it was unfair to discomfort the whole system due to a local failure, for which we might have played a role as managers undoubtedly, but so far, we were blind about the circumstances of the event. We intended to push the boundaries of (dis)comfort until someone "breaks" and explains what happened. Is this an ethical way to manage? Maybe yes, maybe not, but this was the solution I devised back then, and it finally worked. I thought of no other option since I had patiently tried all "soft" paths to persuade the persons involved to speak out.

One afternoon, about ten days after the event, a non-commissioned officer came to my office and started crying, apologising, and explaining the event. He was one of the persons on board the vehicle when the event happened, but not the one driving or the one "in charge" in military terms. The junior officer on board was the driver. It was a team of three who had driven the vehicle to the central maintenance squadron of our base to transfer some spare parts. They were in a hurry to get back to our squadron, change their clothes, and get the free bus back to their homes. At a specific location, the junior officer-driver was supposed to lead the vehicle through a clear and visible small path, but she started driving towards the corner of a semi-sheltered car parking place. The other two persons on board alerted the officer that the vehicle's back cabin was too high to fit safely under the metal beams of the structure. However, she ignored them, and stuck to her decision to drive towards the corner of the parking place; this would save time, which retrospectively speaking was only two to three seconds. I am not judging here the perception and decision of the junior officer, but I am trying to explain to the reader the whole context. As you can guess, the cabin did not fit, and that was why it was malformed and bended. According to the account of the reporter, the junior officer made a few phone calls, and she commanded the team to keep the event secret. They returned the vehicle to the squadron when almost everyone was about to leave for home and they got the free bus without reporting the event to anyone.

After calming the non-commissioned officer down and getting the information I needed, the next challenge was how to approach the junior officer without exposing the person who disclosed the course of the incident. I did not want to damage collegial relationships, but I was impatient to listen to the account of the junior officer and then allow the system to return to the "as usual" state and give back the "privileges" lifted. I made up a story and sent a text message to the junior officer: someone from the maintenance squadron had witnessed the event and the passengers of the vehicle and phoned me to ask whether we managed to fix the damaged vehicle. She texted me back with a long apologetic message. I suggested her to meet and discuss in person the next day.

We talked at my office; she looked nervous and sad. I tried to calm her down and ensure her of my good intentions. Regardless of my assertions, I was aware that she was afraid; I was her chief engineer and a senior officer, and she was a young officer. To be honest, my interest shifted from understanding the unfolding of the event to finding out why she had not come to me voluntarily and earlier to discuss. I was still feeling uncomfortable with the idea that I was not trusted; my ego had been hurt.

The event's description was the same, but I learned more about the underlying factors of the event itself and the secrecy around it. The key point to the latter was that the junior officer had phoned her father, who advised her to shut their mouths firmly and not disclose anything because their career could be harmed, especially the career of hers as an officer and the person in charge. I was not surprised; the junior officer's father and I shared the same impression about the system. The difference though was that I was representing the system before the eyes of my staff, and I felt inconvenienced because I had not been trusted. It was strange to realise that I was not just an individual senior officer reporting to the system, but a reflection of the system at the same time.

Our discussion lasted for about an hour; during this time, the junior officer calmed down, and we agreed that we should draft a plan to restore trust and ensure fairness. Once more, considering the military context and the hierarchical distance between us, I am not sure whether it was an agreement and not just her endorsement to my suggestions due to the hard situation in which she was. I hope the former, but I cannot verify it. The plan was the following:

1. All persons on board the vehicle during the event would transfer it to the vehicle maintenance workshop (i.e., at another squadron) for repairs, and they would help the technicians there to bring it back to an acceptable condition. I phoned the commander of that squadron, pleaded with him to avoid reporting the damage up to the base level, and promised that my technicians would help with the repairs as a means to relieve his staff from the "unreasonable" extra burden. He agreed.

2. The junior officer would prepare a presentation on behalf of all persons involved to explain the event sequence, their team interactions, the decision-making and the reasons of silence, and communicate their reflections during our weekly meeting with all ground staff (note: this

is a strictly internal meeting). The goal was to share knowledge, experiences, aftermaths, and trigger thinking and discussions amongst all technicians.

3. For a given period, the junior officer would be deprived of her "privileges" and transfer them to other persons who suffered the consequences of lifted "privileges" and strict controls. That was somewhat an active apology to her colleagues.

Everything was executed as planned and everyone appreciated the solutions. Lessons were learned by the ones involved, the ones who listened, and the staff involved in restoring the system technically and trustfully. The intervention of the base staff was avoided, nobody was accused, the vehicle was back in service, and the "restoration" and "communication" principles were followed afterwards in all similar events (i.e., unpleasant events never stop occurring) and accepted by everyone. After a few months, my staff reported that they felt more comfortable putting their efforts into restoring the system technically, and less comfortable presenting their stories openly and reflecting. However, despite the justified unease, everyone recognised the value of reflections for our improvement and growth.

Was there something more I could do or anything I should not have done? I cannot say, because back then, I was functioning under multiple cognitive and emotional pressures, resource constraints, and limited time. Would I follow the same approach today? I cannot say too because I have been exposed to many more situations, knowledge, and paradigms since then. Do I regret that I was incompliant and did not report anything to the base? Definitely not because we managed the problem internally and successfully.

My main message to the wider professional and academic community is: when the wider system is perceived as unfair, its subsystems might devise ways to ensure internal fairness as long as they have the support from their managers. The long-term downside is that the wider system remains ignorant, but also calm, and the subsystems further lose their trust in the former.

Even after years following this event, I maintain regular personal contact with the junior officer and many of the technicians I worked with back then; we meet when possible, I play with their little angels, and we look back to those years with nostalgia. This is another sign of success; positive relationships that last years after having been confronted with hard moments.

Keeping Colleagues Happy, but Compromising Safety

It seemed to be a normal, almost perfect, evening; all aircraft ready for their night training flights, all technical forms signed off, and all technicians rested and available to support the flights. My comfort and perception of reality did not last for too long. While on the tarmac, just after the first formation of two aircraft had taken off, the maintenance officer on duty informed me that one of the pilots asked to see me at the aircraft shelter where he had gone to prepare for his night flight. Shortly after, I received this request through radio. The technician who had most recently inspected the particular aircraft approached me and informed me that one of the main wheels had impacts which were out-of-limits according to the maintenance manual. However, he had signed off the aircraft form as "airworthy" and instructed the junior staff to roll the aircraft to a position where the problem would be invisible to the pilot.

The inspector's decision and instructions were driven by the lack of an immediately available spare wheel when the problem was discovered by the junior technician and his intention to avoid delays and, possibly, cancellations in the scheduled flight operations. The squadron commander had informed us earlier that day that the specific training mission was highly important because it was in collaboration with aircraft from other bases. Any cancellation would mess the overall planning and develop perceptions that our squadron was an "unreliable" partner. Therefore, everyone should try her and his best. Admittedly, the squadron commander was a strong advocator of safety; but it seems that his good intentions to accomplish the mission and the transfer of his expectations to junior staff functioned differently. The "best" overrode the "safe" as the reader will understand below.

Back to the scene. I was astonished when the inspector informed me about the technical defect. How many "out-of-limit" issues have been masked in the past and may be covered up in the future without me knowing? I was holding a huge portion of responsibility for the safety of people, aircraft, and equipment as the chief engineer of the squadron. Ignorance about safety-critical issues for which you are co-responsible and co-accountable is not a pleasant discovery. The "bad luck" for the inspector at the specific case was that the junior staff did not successfully mask the problem, and the pilot noticed it; pilots are knowledgeable, trained, and capable of spotting major technical abnormalities. The pilot, naturally, got worried and called for the squadron commander, who, in turn, called for me through the officer on duty. Therefore, five people ended up in the shelter looking at the problem and ready to discuss it: the pilot, the junior technician, the inspector, the squadron commander, and myself, the chief engineer. Actually, after verifying that there was a problem, everyone was looking at me and waiting for my evaluation.

No doubt, the impacts on the wheel were beyond the acceptable limits. Actually, the impacts were not deep into the wheel, but yet quite extensive on its surface. The inspector broke the silence; this exact shape of the defect was what comforted him, and he decided to sign off the forms and release the aircraft for the flight. While waiting for the aircraft to land after the mission, his team would have gathered the spare parts and tools necessary to replace the defected wheel after landing.

Given the circumstances above, it was me now who had to express the final opinion about the airworthiness of the aircraft and advise the pilot and the commander. No time to perform a SWOT analysis, make Fault and Event Trees, consult risk matrices, registers, calculate exposures, probabilities, severities, and all the technicalities we employ when we perform desk-based risk assessments. I had to decide in a few seconds whether the flight should be cancelled or not. What were the implications of each option as I perceived them back then under the conditions explained above?

The safest and most compliant option would be to ground the aircraft and cancel the flight; this would not be the first time, and the pilots and commander had never complained to me about cancellations when there were technical problems and concerns. But, as outlined above, this was a special collaborative mission, and the commander was very interested in its accomplishment and success; otherwise, both the pilot and the commander had the authority to ground the aircraft. Why were they asking for my opinion about the defect? It seemed like they were hoping that I would release the aircraft.

Strange, isn't it when looking retrospectively? How come that the success of a training mission took priority over safe-guarding procedures? However, any expected rational outcome in this particular case was firmly beaten by emotions; what the base commander, the commanders of other bases and squadrons, and other aircraft crew participating in the mission would think about us if we would not have made it? Pilots and technicians in these cases share the successes and the failures, the congratulations and disgrace. The failure to join an important mission, as important as it was perceived at least, was a matter not only of the pilot or the technical teams: it was about harming the prestige and reputation of our squadron. Inversely, any success would lift the latter.

But it was not only about the expectations of the commander and the pilot, who were rather reflecting the expectations of the whole squadron as described above. Grounding the aircraft would immediately mean distrust in the inspector who had released it, and disapproval of his work; it would signal a misbelief in his skills, competencies, and technical judgement. He was a very experienced technician and instructor, a dedicated employee, and in plenty of cases in the past, he had sacrificed his free time to stay longer at work and put in genuine efforts to support our maintenance activities. He was expecting from me to show trust in his evaluation of the situation and support his decision to characterise the aircraft airworthy under a waiver of replacing the damaged wheel after this flight. If not, the message that I was not confident in my staff would spread like an epidemic across everyone within and outside the squadron. How could I lead and manage a team of 100 people who think that I might not be confident in their skills and judgements and stand by their decisions? Whatever their mistakes, they expect visible support from their leader, justifiably or not. The rational decision-making in this case too is beaten by emotions. Was I ready to carry the burden of disgrace from below?

The option of releasing the aircraft definitely meant that I was ready to accept the personal consequences of an incident, but this was not my only fear. When adverse safety events occur, regardless of the people involved in the decision-making and actions, a heavy blanket of mistrust could shade the subsystem – the squadron. How could I face again the people who had trusted my engineering judgement, junior and senior staff? This, now, was a blend of rational and emotional thinking. But the daemon of probabilities came to my ears and whispered: you have taken technical risks in the past, and it worked fine; why is this different? Every case is indeed different, but my rationalism did not win.

I finally released the aircraft; retrospectively looking, it was an unfortunate decision, and I think this was a risky and unsafe judgment. Yes, after the aircraft took off, we started planning with the inspector and the team our actions if any problem occurred during or after the landing. We had our emergency response platform ready, all tools and equipment tested, the vehicle engines on, and our ears and eyes wide open to listen out for and observe the tiniest of the signals that something was not going right. But I was so nervous; this was not what we call proactiveness, and this was not aligned with my professional and moral principles. The aircraft landed safely without any eventuality. It was only God that I thanked; all the rest had failed despite the positive outcome.

How did I, a safety professional who had investigated incidents and accidents of others and spotted their faulty risk assessments, accept this risk? I am still wondering. I preferred to meet temporary, and unimportant in hindsight, operational and personal expectations; I was afraid to deal with personal consequences of possible negative attitudes of my subordinates towards me if I had cancelled the flight and disapproved the judgement of the experienced inspector. Not wise indeed, a huge failure of mine undoubtedly. Lessons learned.

My message to the reader: whatever the training you follow and offer to others, the formal approaches, tools, techniques, and other means to assess safety risks are based on the assumption of rationalism and sufficient resources (knowledge, time, and material). However, many safety-critical decisions are made in the field where the preconditions for a rational risk evaluation do not always exist, while, at the same time, there are factors, such as emotions and relationships, that cannot be inserted in the risk equation beforehand. Even approaches to time-critical decision-making training and education communicate a process-like concept without considering the multitude of interactions occurring within and outside the mind of the decision-maker. The suggested solution? Trigger yourself and others to reflect on their decisions and not just describing them alone; schedule workshops and round tables where people can share their good and bad stories openly and focus on the "why" with an inwards direction in addition to looking around for external factors and influences.

Chapter 14

Are the Stakeholders on Your Side or Not?

Nikolaos Gkionis

Contents

Both of my stories below reflect a role at Public Safety and Emergency planning in a group of companies used to develop and operate mix-used destinations with public areas being a big part of their realm. The stories show that having the critical support of an accountable and leadership-driven top management who can genuinely think out of the "Safe Standards" box can make a big difference or not in the field of safety, where the wellbeing of humans shall always be a high priority.

Pedestrian Safety in Globalised Urban Environments: Waiting for an Accident to Occur

My failure case refers to a sequence of near-miss incidents of similar nature, related to pedestrian safety in fast-growing globalised urban cities. It argues how the failure to approach this very sensitive issue with common sense-making and a good understanding of the environment, context and the limitations and constraints we operate in finally resulted in a serious accident. During this period, I had a very challenging role to lead the establishment, development and implementation of Public Safety Risk and Emergency Governance across all the company's destinations and assets. And I say challenging mainly because of two reasons.

First, the company was in a fast-growing mode, and there was a rush to deliver and operate projects without a proper focus to a Risk Management strategy to allow embedding all adequate safety features within the assets' designs. To me, this approach seemed irrational when the size and complexity of the projects necessitated a more consistent risk management process, especially during the last stages and handover of the project deliverables. Briefly speaking, the delivery of projects within timelines was the highest priority and the most influential driver for the business, and, consequently, the case I describe below.

Second, the company used to operate in one of the so-called Future/Smart cities where the signs of globalisation and urbanisation were apparent in all their possible manifestations, with a big number and a diversity of nationalities as one of the main characteristics and another major influencer in this specific case.

Back to the story now. Since the company dealt with large urban projects, the majority of its assets' layouts were designed to have a lot of open public areas, promenades and walkways with enough capacity to host large numbers of pedestrians. The problem is that, in some of the assets, these public areas were located along extremely busy and high-speed roads, and their design included no physical barrier protections or redundancies between the public areas and roads (e.g., bollards, medians). Thus, the roads were attached to the walkways, a situation that perhaps in other cases, might seem rational and acceptable as, for example, in countries where driving culture and risk perception become embedded early in the lives of their citizens (e.g., education) and supported by rules and regulations which are consistently respected. However, the city environment outlined above suggested otherwise in our case as pedestrians were exposed to high risks, something that nobody had realised. Unfortunately, three consecutive near-misses in a short time, and, finally, a serious accident proved it the hard way.

When I received the first near-miss report, I started recalling memories from my personal experiences as a driver across the roads of the said region. Considering myself as very risk-sensitive in my personal life and specifically regarding road safety, through my first months of driving I felt that this city was not particularly the safest place in the world to drive or walk. There was a perfect mixture of adverse factors out there, waiting for an adverse event to occur: mixed nationality residents coming from countries with completely diverse driving capabilities and habits, which, to my view were often very poor; drivers showing signs of inadequate to inexistent control of visible risks; traffic rules and regulations either not followed by drivers or not particularly focusing on the main causes and factors of road safety; and a stressful urban environment within which citizens prioritised productivity and possibly ranked safety issues low in their list of concerns.

The bad news for me upon the reception of the report was that all these macro-level factors that I had realised, and the accumulated risk they were introducing, had now been transferred to my realm of responsibility for managing the generated impact and consequences. As a company, we had a duty of care towards the safety and welfare of our patrons even if we had nothing to do with the root and underlying causes. When I saw the video footage and pictures from the incident report, I couldn't believe my eyes and realise the extent. There was a car which, after a right turn, rammed onto the lawn promenade and continued its route uncontrolled onto the sidewalk for approximately 50 m before stopping at a bench only a few meters away from the guests of a coffee shop. When I visited the site to investigate further the circumstances of the incident, I additionally realised that the uncontrolled route had started immediately after an inner turn following the crossing of the traffic lights, which expectedly should have captured the driver's attention so that he reduced his speed. But none of these happened. The loss of control from the driver and the route followed couldn't be rationally explained considering the dense city environment and busy road, the high-peak traffic hours and a few more warning factors which could have alerted him (e.g., lighting, vehicle tremors and noise from the small crashes with obstacles during the route [small trees and plants]).

Furthermore, the authorities after investigations and for security reasons couldn't disclose all the necessary data of this sensitive nature incident. Therefore, I couldn't get in my hands any information related to the physical/mental condition of the driver or possible malfunctions of the vehicle, which might have contributed to the loss of control. But to be honest, I didn't need this information. From my point of view, I should not focus on this incident alone from a microenvironmental perspective; instead, I wanted to employ a macroenvironmental approach and look at the poor driving regime in the city and the risks which could generate within our assets' realm.

I considered this first near-miss as a concerning issue with high potential to create a significant risk exposure, and I started analysing data to detect contributory factors as a means to propose timely corrective actions and mitigations to Asset Management and the Senior Corporate Management. It was apparent that there were few more assets with the same design and layout, under similar environmental circumstances and, subsequently, under the same risk. Hence, it seemed to be a good opportunity to treat this risk comprehensively and collectively for all assets exposed. Unfortunately, for the majority of the key stakeholders, this first incident was just an isolated case which could happen to us as happens all over the world. My efforts to raise a warning flag and advise them that we may have more incidents like this in the future were in vain.

Sadly, I was proved right. In less than a month, two similar near-miss incidents occurred at other assets with similar design characteristics. Again, in both incidents, there was a loss of control of the vehicles that rammed onto the sidewalks and crushed into poles and benches respectively, only a few meters away from pedestrians. The unique feature of all three cases was that they did not occur in areas where there are greater risks due to possible conflicts between vehicles and pedestrians (e.g., zebra crossings, traffic lights, safety islands or medians). On the contrary, in all these incidents, the vehicles after losing control literally rammed onto the pavement and sidewalks where unaware pedestrians were walking. By watching the video footage, someone would easily collate these incidents with recent terrorist vehicle ramming attacks across Europe. Moreover, the near-misses occurred at different times of the day (most of them in the morning, afternoon or late night hours), involved drivers with different genders and age (more often young men at their 20s, ladies in their 40s and men in their 50s), different vehicle categories (e.g., sports cars, small trucks, normal sedans) and took place under different traffic circumstances. All these different facts implied that there was not a specific trend or pattern that could potentially be isolated and treated, but rather a result of a driving regime.

From my point of view, it was clear that the macroenvironment factors were the ones we should search to detect deeper causes, like the ones following from personal observations and the messages included in the campaigns of local authorities:

1. Distractions while driving due to the excessive use of mobile devices. This is a prevalent habit in this region despite numerous efforts of the authorities to address it.

2. Poor driving capabilities and unsafe driving practices arising from a highly diversified population of drivers. The vast majority of the population, and hence the drivers, comes from countries with less focus on proper driving behaviours and they have difficulties in adapting to their new globalised environment.
3. Excessive and irrational over speeding considering the road and traffic circumstances at each time.
4. Risky behaviours and attitudes of pedestrians and drivers in areas of conflict as indicated by a lack of mutual respect and an emphasis on individual priorities.
5. Accumulated fatigue for taxi, bus and truck drivers due to occasionally excessive hours on shift.
6. Lack of adequate child supervision.
7. Distracted pedestrians due to talking on mobile phones while walking, leading to loss of contact with the environment.
8. Traffic rules and regulations either not followed by the drivers or not reflecting and adequately enforcing the principal causes of the problem, as outlined in the points above, but limited to fines rather than a proactive presence in the field.
9. High-traffic vehicular load at the roads across all the company's assets.

Apparently, all the factors above regard areas where only the governmental footprint could change through focused and enforced laws, education to influence attitudes and culture and the establishment of engineering controls. The only factor we could possibly directly manage, and therefore take some mitigating measures, was the isolation of pedestrians from road traffic. We should create an enhanced perimeter with a series of safety features in the design to not allow vehicles entering pedestrian areas. The goal was to find a way to "contain" the human error which we accepted was a given reality. After the second and third incidents, Asset Management and C-Suite (i.e., the executive-level managers within the company) started showing concern about the matter and requested from me a full analysis. I studied the incidents as I did for the first near-miss by analysing the proximal and underlying causes; I was aware that my analysis could be probably misinterpreted or not well-received because I was referring to macroenvironmental factors.

Thereafter, I examined and proposed a series of measures to C-Suite to safeguard these areas from our side of responsibility (crash rated bollards, warning signage, speed humps etc.). After the full analysis and assessment were submitted and reviewed, Asset Management and C-Suite described my recommendations as overwhelming. Actually, they looked mainly at the quite big capital expenditure that should be allocated for this scope rather than understanding the magnitude of the problem itself. Moreover, to safeguard their opinion, they claimed that as long as the regulatory authority had not recognised any issue at these areas and had not implemented any additional safety measures, the risk was at an acceptable level and there was no need for further interventions. This was the time I started becoming worried. I couldn't consent to these claims and the company should not simply accept that there was no concern. My next move was to request a joint meeting with the relevant authorities and all the relevant parties to deal with the problem and examine the available options collectively.

It took some time, but the meeting did eventually take place. From the company's side, we advised the authorities about the existing problem in the specific areas, and we highlighted the associated impact in our assets, although we had no control and authority over most of the causes generating the particular risk. Moreover, we presented the analysis of the three incidents as well as the mitigation measures which, from our standpoint, could reduce the risk. Unfortunately, the authorities didn't really show a complete understanding of the risk level, and they asserted that all relevant safety standards had been implemented at the roads next to pedestrian areas. Also, they stated that these kinds of incidents are sometimes accepted as inevitable and could happen anywhere as a result of the unpredicted human factor. To my view, it was complacency and a "Standards" norm that didn't allow them to demonstrate the appropriate willingness and comprehend the real uniqueness of these cases within their context, meaning the particular urban environment and different circumstances. The discussions during the meeting made me realise the presence of a silo-driven approach to risk, the establishment of a belief that the implementation of standards can minimise risks everywhere, and an attitude that customisation of risk measures to different environments was unnecessary. All these were the perfect ingredients of a recipe to inhibit the investment in additional and solid solutions. Subsequently, C-suite was now backed by the statements of the official authorities; on top of that, these statements functioned as the ideal justification not to proceed with our mitigating measures. Complacency replaced awareness after all.

Unfortunately, once again, humans had to learn the hard way in the aftermath. Only after a month, a serious accident happened in exactly the same way; the additional element was the emergence of back and forth blame among stakeholders, which is a typical phenomenon is such serious cases. All the parties involved in the meeting outlined above, during which they had concluded there was no need for interventions, were now in a big rush. They started chasing my team and me to meet them urgently, finalise as soon as possible my proposals and proceed with fast-track budgetary arrangements to put in place whatever was necessary to avoid such accidents in the future. For me, it was a systemic failure. We had to wait for the accident to accept the reasons, which had been recognised long before, and, magically, find money to invest in minimising risks.

The lesson learned for my team and me was that this unfortunate example could be used in the future to stimulate how important it is to include and embed safety aspects at the early stages of the design and development of projects. There is a need for proper studies to take under consideration not only micro- but also macroenvironmental factors, which will enable having all customised safety measures and redundancies for the delivery of a project, without the need of back and forth actions trying to explain and justify additional capital expenses necessary to apply safety. The Global Status Report on Road Safety from the World Health Organization[1] shows that nearly half of those killed in road traffic accidents are among vulnerable road users with 26% representing pedestrians. As per the specific report, there is a dramatic growth in the numbers of motor vehicles and the frequency of their use around the world, but the most worrying part is how negligent and sloppy drivers are nowadays and how vulnerable pedestrians have become. From my point of view, this kind of incident has started to create a trend worldwide which will be on the front page in the future. Megacities and urban environments especially offer the perfect mix of ingredients to create conditions for more incidents with increased severity. Pedestrian collisions should not be accepted as inevitable because, to a large extent, they are predictable and preventable if their contexts are understood and their causes are proactively addressed.

Kids' Playground Safety: Compliance with Global Standards Might Not Be Enough

The success safety story regards our perception and apprehension of safety at playgrounds for kids. As mentioned in my previous story, my company's assets comprised a large number of open public areas, many of which included playgrounds with various activities for kids. These areas are operated either directly or through a third party, but all their structures and features are designed, manufactured and built mainly from suppliers outside the region. My story has to do with a newly introduced playing structure manufactured by a renowned company with global market penetration. The specific composite playing structure was manufactured predominantly by wood and was large enough to incorporate many different challenging activities for the kids like balancing, jumping, hanging, climbing etc., thus creating an interactive environment for the kids, but without a specific and formalised pattern and flow of play, a factor which proved important to the assessment and decision process later on.

At the time, the supplier installed the specific structure in our premises, it was compliant with all global safety standards for playgrounds (e.g., BS EN 1176, 1177 and 7188). However, is compliance with these standards alone enough to ensure safety for the children's population in the specific region or should we take additional protective measures? Kids are humans and as such differ on how they behave, play and interact depending on their habits and the degree to which they carry their parent's mentality, behaviour and risk apprehension. The previous experience of Asset Management with playground areas at the specific asset was somewhat worrying. A lot of minor and moderate injuries had been witnessed and reported even at small, ordinary and not particularly hazardous structures. Following investigations, these events had been attributed more to inadequate supervision and a poor risk apprehension rather than risks inherent in the design or functionality of the respective structures. Consequently, when the aforementioned new play structure was installed, Asset Management, having these near incidents in their record, expressed concerns about the level of safety for the kids and was hesitant to grant their permission to open the area for the kids. Thereafter, they decided to keep it fenced until further notice.

Someone could claim that a quite big portion of incidents in playgrounds can be accepted or tolerated because they are places of fun and excitement with massive amounts of energy, mobility and velocity. However, we should not forget that the challenging activities these places offer come with associated risks, which, when considering the unpredictable behaviour of kids and their naturally immature risk perception, can generate unforeseeable negative results. Furthermore, many of us might have experienced times when even custodians demonstrate complacency towards their kids' activities (e.g., kids are kids and it is expected to get injured while playing). However, when unfavourable events take place, exactly the same people start pointing their fingers merciless against everyone involved. Hence, apart from kids' safety which is the ultimate goal, any undesirable event can generate signifi-cant risks for the company at reputational, financial, legal and social levels. In such cases, neither any form of signed disclaimer nor a third party liabil-ity insurance are able to protect the company and commensurate losses.

Shortly after the decision of Asset Management, I was informed about their concerns and asked by them to conduct an inspection further to the risk assessment statement the supplier had submitted. First, I viewed images of the structure and realised that the particular structure would be very dif-ficult to operate and host kids safely in the specific environment. To verify my initial observations, I visited the site having at the back of my mind all the historical data of frequent injuries at the other play areas. My visit con-firmed the concerns of Asset Management and mine about the opening of this function, as I identified significant hazards and inherent risks associated with the interaction between the design and the described kids' activities in the manual. I explain more about these hazards and risks below, where I list the results of my risk assessment to gain an understanding of the context and collect the information needed to educate the relevant parties. In the meantime, I had to convince all stakeholders about the reasons we considered the opening of this specific play area unsafe as well as the next steps.

I requested the supplier to provide the respective manual along with the play value analysis to assist me in developing a complete understanding of the activities. Play value analysis shows the added benefit of a challenging activity or interactions of activities to the mental and physical development as well as motoric competences and social cooperation of kids. I exploited these pieces of information along with the data and experiences from past incidents at our playgrounds and my risk assessment which was based on the interplay between the design of the structure and the described activi-ties presented to me in demo videos from similar playgrounds by the sup-plier since I didn't have the ability to assess them in real-time. Thereafter, I managed to create a clearer picture concluded to the issues described below.

Playground Design Safety. The design of the structure didn't restrict the various movements and playing activities into designated paths/ routes (e.g., the same ladder would be used from kids to climb up and climb down simultaneously). Also, the space between the different components of the structure was very limited to be considered safe, with only small areas for moving, jumping, running, hiding, climbing and, generally, accommodating the typical rapid reactions of kids. Furthermore, the material used for the structure was considered hard and rough; thus, in many areas where a loss of balance could happen (e.g., running, pushing, jumping), any impact with head parts could cause serious injuries. Additionally, a few platforms allowed fall from heights that exceeded standards and could lead to serious injuries regardless of the bottom surface material. Moreover, guardrails and protective barriers were not in the right place for elevated surfaces, including platforms and ramps. Last, there were many areas identified with the potential of entrapment.

Unregulated Activity. There were no particular play rules or separation tactics to regulate the different activities within the structure. This combined with the fact that each activity didn't have a formal pattern or flow of playing, as mentioned below, meant that any activity of one kid could pose a risk to the neighbouring activity undertaken by another kid in a circular pattern. Also, there were activities within the structure aimed to completely different groups of ages and children's heights without any separation strategy to keep smaller children away from areas designated for older kids and vice versa. I want to note here that according to the manual provided, the structure offered activities for ages ranging from three-plus to ten-plus without this being mentioned on the signboard. Briefly speaking, any kid could access everything within the structure regardless of age and height. Furthermore, there was no requirement for trained and competent supervisors to guide and monitor the activities.

Behavioural aspects. Based on our experience and information from different incidents under various conditions, we believed that kids' supervision was really poor, if existent at all. Custodians, either parents or nannies, due to different reasons, which we thought they were linked to cultural and habitual factors, had rarely shown active interest or put limits to kids' activities. Certainly, we cannot exclude here the perception that play structures have been approved, meaning that all major risks have been managed and, consequently, the custodians felt they were entitled to enjoy some time free from worries. However, considering that all these active and spontaneous children might not have been familiar with complex playing structures, we believed that there were significant risks. Kids at playground areas are unaware of their own limitations, and they rarely have the necessary experience to appreciate risks. Hence, taking into account the factors of flawed design and unregulated activities stated above, adequate adult supervision should be an integral part of providing a safe environment.

Following the observations mentioned above, I had in my hands a set of solid reasons to justify the suspension of the structure, and I asked the supplier to offer his explanation and points of view on the results of my risk assessment. I also suggested that Asset Management request from the supplier to submit to us all the documentation supporting their risk assessment and compliance with the standards and schedule a meeting to simulate the activities point by point. The supplier provided documentation which applied to the country where the structure was produced and referred to a very generic risk assessment for the overall structure without any reference to the whole range of activities or necessary customisation to factors and conditions of each region.

After the meeting with the supplier and the submission of our report to all parties, nobody defended the release of the structure, and after approximately a month, I was informed that it was removed from the asset. To be honest, I was really surprised, and I felt very happy when I heard that. I didn't expect that outcome even though I was confident in my opinions and proposals. One reason for doubting about any favourable outcome was the positive reputation of the supplier globally. Another reason was the possible reluctance of management to reject the project and make the project team feel "defeated". It was a big win for the public and kids' safety, the department, me and the company alike. Even more importantly, it was a sign that a positive safety culture started being developed with this asset being the example case and lighthouse for future similar cases. Having asset management on my side, and not against me as happened in the previous story where commercial interests overrode safety, made my position stronger and it was much easier for me to deal with the relevant stakeholders. Furthermore, I was happy to realise that this demonstration of interest in this case was purely ethical, justifying at all grounds the "duty of care" of the employer.

My key message to the reader is that, indeed, there are standards with which we must comply. However, we must understand that standards suggest the bare minimum; different cultures, diverse perceptions and interpretations of hazards, local customs and habits and other factors can render an acceptable risk in one part of the world a "stop activity" somewhere else and vice versa. If we approach risks by gaining a deep knowledge of the environment in which we operate and engage with humans, infrastructure and technology, we will become more capable of avoiding devastating consequences.

Note

1. https://www.who.int/violence_injury_prevention/road_safety_status/report/en/

Chapter 15

Making Safety a Priority

Sikder Mohammad Tawhidul Hasan and
Mohammad Tahidul Islam

Contents

The nation was taken aback with two devastating accidents in the ready-made garment industry of Bangladesh in 2012 and 2013. In 2012, a fire blazed at Tazreen Fashion and left the factory in ashes, claiming 112 lives and leaving more than 200 people injured. Just over a year later, not only Bangladesh but the whole world was shocked by one of the most unanticipated catastrophes in the world with the Rana Plaza collapse, which resulted in 1134 fatalities and about 2500 injured workers.

A retrospective look at those accidents indicated that they were the aftermaths of non-compliance and infringement of national safety laws and rules. Factory owners tended not to abide by regulations. For both cases, the factory owners violated the Bangladesh National Building Code (BNBC) and Bangladesh Labour Act (BLA), 2006. Regarding the Rana Plaza case, the factory owner had permission to construct a six-storey building; however, BNBC was violated with the addition of three additional floors. The code was also broken by not keeping the electricity generators on the ground floor. In the case of Tazreen Fashion, the main exit was blocked, thus breaching the respective BNBC guideline. In both cases, the factory owners were aware of the applicable rules and regulations and violated them purposely.[1,2,3,4]

Bangladesh is the second-largest ready-made garment exporting nation in the world, serving more than 200 European retail brands and about 28 North American brands in the US and Canada. Nevertheless, before these catastrophes, the tripartite stakeholders, namely the government, employers and workers, were not fully aware of the importance of Occupational Health and Safety (OHS) in addition to any codes of practice and labour-related legislation. From the government side, no National OHS day was observed contrary to global practice where such campaigns aim to bring all stakeholders under one umbrella and a platform to convey critical messages about OHS issues.

Following the Rana Plaza collapse, due to concerns over the labour market reformation and non-compliance to safety regulations, Bangladesh was denied benefits from the Generalised System of Preference (GSP) which facilitates duty-free access to the US. Although ready-made garment products do not fall under the GSP system, the collapse damaged the reputation of the entire sector. After that accident, immense pressure was imposed by different stakeholders around the world to improve the working conditions of ready-made garment factories with a focus on safety. Stakeholders included international buyer communities such as Accord and Alliance that represent European and North American buyer associations.

Therefore, after those fatal events and the increasing pressure imposed by the international buyers association, the tripartite stakeholders eventually acknowledged the significance of OHS, and, in view of this, the government took initiatives to strengthen and reform the labour inspection department, amend laws, introduce new rules and policies and develop a draft of the National OHS action plan as one of the essential agenda items of Sustainable Development Goals (SDG) under SDG-8 (Decent Workplace and Economic Growth). Briefly, the government of Bangladesh had to take rigorous measures to regain and further uplift their damaged reputation in the international arena. The recruitment of safety inspectors was one of these initiatives.

In the frame of the above, in 2014, the Department of Inspection for Factories and Establishments (DIFE) was transformed from a directorate into a department with a massive reformation in terms of uplifting the rank of the head of the department, recruitment of new inspectors, establishment of district offices and ties with international stakeholders. After restructuring, the inspections started with the preliminary assessment of the ready-made garment factories with the collaboration of the DIFE, representing the government end, and the International Labour Organization (ILO) under a project named "ILO Programme on Improving Working Conditions in the Ready-Made Garment Sector in Bangladesh". The project was funded by Canada, the Netherlands and the United Kingdom and was launched in October 2013. The assessment was conducted based on actions identified in the National Tripartite Plan of Action focusing on fire safety and building integrity.

A Safe Workplace Is a Life-Saving Investment: The Earlier, the Better

Under the conditions described above, we started our journey as safety inspectors at the DIFE. Our responsibility was to assess and detect whether the factories inspected had any structural, electrical and fire-related issues. However, we were not equipped with the required technical equipment to conduct the inspections properly and detect all possible deviations in the factories. This happened due to budget constraints and management's lack of context knowledge and unawareness of the types of equipment required for an adequate inspection. The new administration in charge needed time to understand the situation immediately after the disastrous Rana Plaza accident amidst pressures from national and international stakeholders. Also, before the DIFE reformation, there were very few safety inspectors; following this significant change, the DIFE was staffed with newly recruited inspectors like us who were not trained for specific safety inspections.

Besides, under the mandate of the project to inspect the whole ready-made garment sector, the number of safety inspectors was not adequate. Most of the inspectors were electrical and mechanical engineers, and only a few of them were civil engineers. Moreover, in our department, we had no fire engineers with a relevant degree in fire engineering. To mitigate this issue, we had to associate with engineers with appropriate expertise working for external firms and institutions such as Tuv Sud Bangladesh Pvt Ltd, Bureau Veritas and BUET. Despite the adversities outlined above, we constructed an inspections guideline for our preliminary assessments by using the BNBC as a basis and engaging DIFE inspectors and management as well as other international stakeholders (e.g., ILO) during several brainstorming sessions. After finishing the guideline, we started the assessments with the help of a checklist developed with the collaboration of the DIFE and the ILO and designed specifically for the ready-made garment industry in Bangladesh.

Over the course of our inspection journey, we were first assigned a few factories for preliminary assessments and most of the inspections had a proactive mindset. Regarding the case of structural evaluation of a factory in the Narayanganj district, the major challenge we faced was to infuse the notion of safety into the factory owner's mind and explain to him the importance of addressing non-compliance issues. At first, the factory owner was reluctant to understand the consequences of deficiencies not being remediated. We will get back to this point after explaining the process and findings from our preliminary assessment of the factory.

The specific factory was a six-storey building with an area of 60,000 sq. ft. When we started our preliminary assessment, we asked the management to show us all structural, electrical and fire-related documents. However, they failed to provide the relevant documents and, therefore, we asked them to provide the complete documentation and drawings based on the estimated time frame according to the National Tripartite Plan of Action (NTPA) guidelines.[5] The specific official publication considered the intensity and priority of the remedies and allowed six weeks for load management (structural), six months for electrical drawings along with electrical load calculations, six months for installing a lightning protection system and six months for installing a fire protection system. During this time frame, members of the review panel inspected the factory regularly to check the progress of the remediation work. The review panel has the authority to make any decision regarding the complete or partial shutdown of any factory based on the remediation progress. The review panel consists of members from the DIFE, BUET, Accord, Alliance and TUV-VEC.

The next step of the assessment was to inspect the roof of the factory building. We found out that the roof was used for different purposes such as a canteen, storage etc. Hence, we asked the factory management to remove all the existing structures from the rooftop because, in case of any fire incident, workers located in other floors might have to reach the roof to seek shelter from the blaze. We also inspected an RCC (Reinforce Concrete) water reservoir which imposed a heavy load on the building, and we asked the factory management to replace it with a plastic water storage tank. Since the safety factor of the building was within the limit, according to BNBC, we did not give any instructions for the immediate suspension of factory operations. Instead, we asked the factory management to arrange a Detailed Engineering Assessment (DEA) by hiring a government-enlisted firm to conduct it.

When we visited the production floor, we noticed that some machinery was on the cantilever portion, which could erode the structural strength of the building and inflict a structural failure. In light of this fact, we asked the management to cease any operations on the cantilever portion and gave them six weeks to dismantle it, as shown in the NTPA guideline. We informed them that we would return for a follow-up inspection to check the progress of the Corrective Action Plan (CAP). We also clarified to management that the DIFE would file a case if the CAP was not implemented within the expected time.

In terms of electrical safety, we found that the factory had a substation inside the premises, but it was not protected. According to the NTPA guideline, we asked them to install a fire-rated floor-to-ceiling enclosure, place an approved fire-rated self-closing swing door for the substation room and use a warning display so that nobody would trespass except authorised personnel. Furthermore, we checked the Distribution Board (DB) with a thermal scanner to measure the temperature, and we found that it was generating excessive heat beyond the expected limit. After consulting with the management regarding the issue, we instructed them to enclose the distribution board with a dead front construction. Regarding the fire safety assessment, we found that there were no fire doors in the factory, and we asked the management to install such doors. Also, we observed that the factory had three lifts, two for staff and one for cargo, which were not fire-separated. We asked them to fix the problem within six months per the CAP provided to them.

After completing the preliminary assessments as described above, we amalgamated all the findings into three separate CAPs, including specific recommendations, and we assigned the factory a particular code of colour which represented its susceptibility status. For structural assessment, the colour code (i.e., green, amber, yellow and red) is based on the factor of safety. For fire and electrical assessment, the colour is based on the time frame (i.e., immediately, six weeks, six months and on-going). After sending the CAP to the factory management with the requirement to act accordingly, we communicated regularly with the factory through emails to track the progress of execution.

We visited the factory to follow-up CAP progress every fortnight. Initially, we found that the management was quite reluctant to proceed with the remediation and was preconceived with the notion that the "cost of a safe workplace is an expenditure, not an investment". Given this, we arranged an awareness-raising meeting with the owners and informed management about the benefits of the remediations by pointing out that "*Everything is uncertain. Anything can happen anytime. If you think that you would do it later and it would be convenient for you, then it is absolutely fine; however, by any time you could be the next victim of any fatal accident. So, why not making your workplace safe now since you have to do this eventually? Always think that it is not only about your life; the lives of workers who are the backbone of your factory are in your hands as well and keeping them safe is your core responsibility, and you cannot just neglect it*".

After the meeting, the owner felt very much motivated, contacted one of the government-enlisted firms and gave them a work order to start a DEA and respective remediations. From our side, we asked the specific firm to prepare all the designs and drawings and submit them to our office. The firm immediately started working on the project while we were continuing our close follow-ups. Within two months, all the drawings and designs were approved and the firm launched the actual remediation works. Eventually, the factory completed all CAPs and received an appreciation certificate from the DIFE.

Our success lies in taking up a mammoth challenge of executing our safety inspection tasks from scratch to improve the working conditions in ready-made garment factories through successful remediation activities. The most crucial aspect of the success was that we managed to motivate the owners by reminding them of their responsibilities for ensuring safety at their workplace. Hopefully, this would help reclaim the positive brand image of the entire ready-made garment sector and benefit the country with regard to economic gain since more than 80% of total export earnings are generated from this sector.

Improvement Opportunities in Failures: A Continuous Journey

As mentioned in our story above, following the Rana Plaza collapse, we started assessing the factories under a nationwide initiative of the government. We were assessing each factory with the use of a checklist the first version of which, retrospectively speaking, did not yield the desired outcome. We felt there was something missing; the checklist did not include any risk ranking and comprised only of binary yes and no options. Also, the questions of the checklist were developed according to the Bangladesh

Labour Law and Rules and consisted of long sentences instead of short, precise items to be checked. We discussed this with DIFE management, raising our concerns that with the current checklist we were not evaluating the actual status of the factory and were not able to assess the extent to which the factory was vulnerable or not. With the support of the ILO and caution to observe Bangladesh Labour Law and Rules, the checklist was revised and explicitly tailored to the ready-made garment industry. As expected, we started using the new checklist during inspections.

However, as safety inspectors, we felt we had failed to provide to the owners of the factories that had been already assessed a comprehensive picture of the standing of the factories in terms of their susceptibility. At the beginning of the project, we could have examined whether this inspection tool was adequate to inform our decision-making at the DIFE and offer meaningful results to factory owners. Life is not black and white most of the time, but our checklist was. In hindsight, we could have invested more time in developing the first checklist, and even pilot it with a few factories to receive comments and revise it. However, the time pressure factor overrode good practice. We wanted to assess the safety levels of our industry, but we started with the wrong tools.

A second systemic failure was the lack of relevant training regarding how the inspection should be conducted, a situation which significantly reduced the quality of inspections in the first year. We, along with other inspectors, were obliged to conduct the inspections under immense pressure from the government after the Rana Plaza incident. Before joining the DIFE, almost none of the inspectors had sufficient, if any, experience in inspecting factories, and there were no Standard Operating Procedures (SOP) for inspections to consult while executing our duties. Thus, during the inspections, we just used the checklist with the YES/NO options based on what we could observe easily in each factory. Consequently, we struggled to detect less visible, but, possibly, safety-critical technical defects due to the lack of relevant training, technical skills and SOPs. Nevertheless, after a year, we started undergoing respective training offered by the government (e.g., the DIFE, Fire Service and Civil Defence [FSCD]) and international stakeholders (e.g., the ILO, Deutsche Gesellschaft für Internationale Zusammenarbeit [GIZ]). Moreover, SOPs were also developed to complement our newly acquired technical skills and conduct adequate inspections.

Furthermore, in our labour law, one of the chapters is about a provident fund. The law describes in detail how the fund should be created and managed. However, the main loophole is that the investment in provident funds it is not mandatory by law. During our inspections, we discovered that, on the one hand, most of the factory managers were not willing to provide the fund, and, on the other hand, they were complaining about high employee turnovers. For any business, considerable staff turnover has negative implications as experienced and trained employees might leave organisations. This means that new employees must be recruited and trained, a situation that costs money and time. Besides, the experience needs time to develop, meaning that productivity can slow down. This, in turn, puts increased work pressure on staff, and, possibly, places safety as a secondary priority.

We tried to find out the real reasons behind the significant employee turnovers. After interviewing several workers, we realised that they were not feeling job security when working in factories. Factory owners could sack employees at any time and the workers would receive no compensation. After having identified this principal demotivating factor that threatened staff loyalty to specific employers for long periods, we collected some data from different factories where the provident fund was provided or not. The analysis of data suggested that factory management who provided a provident fund were retaining employees for a longer time than the ones who did not invest in the fund. We delivered a presentation to the management of a specific factory and shared with them all the information we had collected.

All the board members of the factory management appreciated our efforts to offer a pragmatic view of the positive effects of the provident fund scheme beyond the obvious ones, meaning the financial security of workers. The factory management promised us they would try their best to start the provident fund. They also acknowledged that the idea was excellent and it would help them to retain their employees. After a few days, we followed-up with the particular factory and felt enormously disappointed to find that the management had not taken any initiatives. The owner thought that investing in the fund would hamper profitability. We realised that the owner considered only the profits and chose to ignore worker benefits despite their initial positive reaction during the presentation. We tried our best, but we failed to convince management about the advantage of the scheme. What we presume we could have done better is bringing factory owners, management and workers together during motivational meetings and could have driven decisions which would reconcile the perspectives of all stakeholders.

Another case regarded a safety campaign. The DIFE and the Ministry of Labour and Employment started observing the National day on OHS from 2015. On the very last occasion, the Government of Bangladesh awarded 24 companies from the apparel, finished leather, jute, pharmaceutical and tea sectors for their good OHS practices. However, we realised that we failed to infuse the core messages about the National OHS day during our inspections. After each inspection, we could have organised informative meetings to convey the significance of the National OHS day, an initiative which eventually could have led the ready-made factory owners and workers becoming aware and interested in good OHS practice.

Last but not least, in a few cases, we inspected more than one factory on the same day, a condition that possibly compromised the quality of our work. Sometimes, it was hard to maintain the schedule of inspecting 25 factories and establishments in a month, and, parallelly, accomplish all the official administrative work and compile our reports to the ministry and the DIFE as requested. The duration of each inspection depends on the size of the factory and it takes ample time when it comes to inspecting large factories. Furthermore, apart from ready-made garment factories, each inspector is obliged to check other types of factories (e.g., plastic and chemical) and establishments. Therefore, when having to inspect a large and a small factory on the same day, it was quite tough to carry out in-depth inspections, meaning that we did not meet the expectations of ours and management for high-quality results. Despite whether there was a packed schedule coordinated by the DIFE, we felt that as responsible inspectors, we should have made sure to organise our time effectively by considering the type and size of factories and raise any concerns to management in a timely manner.

There is a saying that "failure is not the opposite of success but a part of the success". Every failure is a learning opportunity and opens the door to ameliorate the situation further. In this section of our chapter, we described a few failures we experienced. It does not matter whether the conditions we confronted stemmed exclusively from us or were the result of broader organisational and national factors. We are part of the system, and we influence it as it affects us. Whatever the negative circumstances and the pressure to perform our duties, as safety professionals, and especially safety inspectors, we must keep aware of the wider context, wonder on our practices and the tools used, reflect on our role in the system and look beyond the bare minimum, namely laws and regulations.

Notes

1. https://www.theguardian.com/world/2013/may/23/bangladesh-factory-collapse-rana-plaza
2. https://www.npr.org/sections/parallels/2017/04/30/525858799/4-years-after-rana-plaza-tragedy-whats-changed-for-bangladeshi-garment-workers
3. https://www.bbc.com/news/world-asia-20522593
4. https://time.com/6607/bangladesh-tazreen-factory-owner-charged/
5. http://rcc.dife.gov.bd/index.php/en/information-resources/ntpa-assessment-guidelines

Chapter 16

The Practical Value of Ensuring Effective Interfaces and Workforce Engagement

Spyridon Markou

Contents

The following stories refer to my recent experience as senior safety manager for a fast-growing private company operating 14 aerodromes in Greece in locations considered as worldwide tourist attractions. I was responsible for the development, maintenance and administration of an effective centralised Safety Management System (SMS), ensuring that risks directly associated with aviation activities and/or those potentially endangering such operations were reduced and controlled to an acceptable level.

Difficult but Not Impossible: Ensuring Safety during Disruptive Activities

Aerodromes are a complex and hazardous environment, especially when civil engineering activities (e.g. works on runways, taxiways, apron) are taking place within airside areas used for aircraft operations. Regardless of the type of activity, such as rehabilitation of surfaces, installation or modernisation of signage (e.g. lights, signs) and new markings (e.g. push-back procedures on the apron), these works often affect normal aerodrome operations and can jeopardise aviation safety. It is not preferable, but, in most of the cases, aerodrome operations and civil engineering activities cannot be disassociated, representing a possibly significant safety risk.

Historically, civil engineering works in airside areas have contributed to accidents or serious incidents worldwide. In the catastrophic collision of a Boeing 747-400 with a construction site at Taipei Taoyuan in October 2000, the aircraft, under poor conditions of visibility, mistakenly selected to take-off from a taxiway in which construction works were in progress instead of using the operative runway. To prevent situations which could lead to catastrophic events, it is of utmost importance that upcoming civil engineering activities are carefully planned, scheduled in coordination with all parties involved and executed in a way that does not compromise flight and ground safety.

When I took over the position of safety manager, my organisation had to balance between the contractual obligations for the renovation of aerodromes within a strict timeframe and the safe continuation of operations, even during the high season. The maintenance of safety levels under such circumstances was considered as an achievable goal only if everyone involved in civil engineering activities was able to identify the emerging safety consequences of their activities, understand their legislative responsibilities associated with their activities and understand the importance of reporting any hazardous situations, incidents and accidents. Under the provisions of the relevant contract, the Contractor was committed to providing all necessary resources and making any reasonably practicable efforts to ensure:

- Compliance with all applicable statutory and regulatory requirements of the State, applicable codes, standards, regulations and resolutions regarding aerodrome operations, and Permit-to-Work System requirements for each specific area and/or activity.
- Coordination with the Safety Department, so that risks associated with planned civil engineering activities would be reduced and controlled at an acceptable level.
- Coordination with the Operation Departments, so that civil engineering activities would minimise disruption of aerodromes operations.

■ Thorough and cautious design and planning of civil engineering activities so that any unreasonably noxious or offensive annoyance or nui-
sance would not be imposed on others or affect any facilities adjacent
to the aerodrome.

To ensure the implementation of the above, I established a process to ensure
that safety requirements were considered before the initiation and during all phases of each construction project. The elements of this process
were Roles and Responsibilities, Project Planning, Safety Risk Management,
Communication, Monitoring/Oversight, Safety Reporting and Training, Safety
Promotion and final Delivery into Operations.

First, I insisted on the clarification of the roles and responsibilities of
individuals implementing the approved planning. I considered this step vital
for the continuous monitoring of compliance with the defined requirements
as well as for instilling necessary vigilance and oversight to detect areas
needing attention in case of altered construction activities. The Contractor
who had the overall responsibility for the administration and timely communication with all parties involved and affected during all phases of the
works nominated a central representative. Moreover, in each aerodrome, a
Coordinator of Works (CoW) was appointed to bear the eminent responsibility for monitoring the progress, ensuring compliance with all standards and
practices, effectively coordinating with the aerodrome operator and reporting any disruption and/or undesirable safety event. In turn, the aerodrome
operator had assigned a supervisor to monitor the implementation of the
agreed mitigation measures and reporting any new or previously unidenti-
fied hazards.

Following the establishment and clarification of roles and responsibilities, it was commonly understood that each aerodrome manager and the
Contractor should conduct regular onsite safety inspections throughout
each project and remedy as soon as possible any deficiencies caused by
negligence, poor oversight or project scope changes. Supplementary to the
Contractor's monitoring actions, the Safety Department conducted additional inspections, observations and audits on a regular and irregular basis
to evaluate the Contractor's compliance as well as the effectiveness of
implemented measures throughout the project and especially before the
deployment of their deliverables into operations. Nonetheless, it was clear
that the oversights executed by the Company did not mean to supersede,
override or take precedence over those of the Contractor, who was holding
the ultimate responsibility for the safety of the works undertaken. In this

way, the Company, on the one hand, was fulfilling its obligations to ensure safety at the aerodromes and being constantly aware of possible work floor problems, and, on the other hand, was supportive of the safety efforts of the Contractor without promoting a culture of complacency that could emerge due to a false sense of protection from the oversight activities of the company.

Project Planning is the most challenging phase in each project's life cycle. Good planning eventually offers an excellent opportunity for all interested parties to express their concerns and any specific requirements, agree on the timeline, costs, resources, quality, change and risk management and, in general, ensure the delivery of outcomes on time and within the predetermined budget. Especially regarding safety risks, we agreed on weekly meetings between the Safety Department, planners and project managers to ensure that the associated documentation (i.e. charts, detailed phasing plan of works and safety risk management) had addressed the organisation's requirements before the commencement and during the execution of works. Also, local meetings were held at aerodromes with upcoming works to confirm that the safety concerns of the employees had been considered. Moreover, we reviewed the Risk Management Report to ensure that all identified hazards and risks had been adequately managed and risk levels of possible consequences had been properly assessed. Equally important, we reviewed the aerodrome Emergency Response Plan (ERP) to ensure that the associated procedures had been amended and confirmed that the aerodrome staff involved, especially fire-fighting personnel, in case of an emergency event, were knowledgeable of the new or changed conditions due to the on-going engineering projects. Last but not least, we organised local kick-off meetings to finalise the Project Planning Phase and ensured that the aerodrome staff community and other involved parties were knowledgeable of the project requirements, including safety matters.

Risk Management was recognised as a core process for an appropriate and comprehensive SMS that maintained and improved operational safety at aerodromes. Risk Management includes the subprocesses of hazard identification, risk assessment, mitigation and control of (residual) risks and continuous monitoring for new or changed hazards and risks. Under this framework and during the Planning Phase, the Contactor was first obliged to develop a Hazard Identification and Risk Assessment (HIRA) register for each project (e.g. rehabilitation of runway surface) and evaluate the consequences (e.g. operational disruptions, safety implications) and risk level of each safety hazard. The Safety Department reviewed the adequacy and completeness of HIRAs according to ICAO Annex 19, ICAO Doc 9859 and the corporate SMS, and, afterwards, the Contractor circulated the reviewed and approved HIRAs to all interested departments as well as the aerodrome in which works were going to take place. Moreover, the Contractor should submit a revised HIRA for review when there were changes to the initial assessment and communicate any concerns as a result of the implemented HIRA.

Apart from the initial and residual risks, we recognised that there were also actual/real risks as a result of the minor day-to-day changes in aerodrome operations, which could not be typically captured and included in the HIRA. To cope with this dynamic environment, we considered as of utmost importance to establish an additional daily brief meeting conducted between the CoW and aerodrome manager before the commencement of any works. This initiative proved extremely valuable as the aerodrome operator was able to have a real view of the progress and potential constraints, assist with required changes and evaluate daily the effectiveness of the safety measures implemented.

Whatever formal management process an organisation introduces, information exchange and real-time feedback are excellent means to deal with current and future challenges and identify trends. Effective communication and timely coordination between all involved parties are the keys to maintaining safe operations during the whole lifecycle of any project, from planning to final delivery to operation. These two factors were even more crucial for my organisation due to its contractual obligations for renovation within a strict timeframe, while the actual construction works were being carried out simultaneously at 14 geographically separated aerodromes. This situation created the necessity for me to establish a communication mechanism for systematic and real-time feedback and information exchange, not only between each aerodrome and our Safety Department, but also amongst the 14 aerodromes.

To facilitate the above, I scheduled bi-weekly teleconferences amongst all aerodromes, during which participants had the opportunity to share information regarding on-going civil engineering activities and raise any significant safety concerns. Lessons were learned, good practices, concerns, suggestions and ideas for the resolution of issues were all shared and comprised the valuable outcomes of these meetings. Also, the meeting minutes were circulated to the participants, so that everyone was aware of the decisions and would be able to refer to them in the future (e.g. similar projects).

Safety reporting can have a direct impact on safety performance. During our training courses, I always pointed out the need for everyone to remain vigilant and report hazardous situations before they escalated to incidents and the necessity to create a culture that promoted protection and prevention instead of learning from events only retrospectively. However, despite the efforts and commitment to implement procedures and measures to reduce safety risks during civil engineering work, when the works began, we noticed that this was not common sense to everyone. We noticed that workers had not been convinced about the importance of reporting, for the reasons explained in the next paragraph. This situation led to the under-reporting of safety events as well as hazards that could potentially lead to undesirable consequences.

Following collaboration with the Contractor and an unofficial survey at aerodromes where construction works were in progress, we concluded that reporting behaviour was negatively influenced by a fear of disciplinary actions (i.e. the belief that all errors are punished), perceived uselessness (i.e. attitudes that management would take no notice to improve conditions), unquestionable acceptance of risk (i.e. beliefs that incidents are part of the job) and impracticality (i.e. an unfriendly and difficult system to submit reports). Thus, it became apparent to us that we should give special attention to these matters. In coordination with the Contractor and the Company's Training Department, we organised lectures and training courses for the Contractor's workers, delivered locally at each aerodrome.

The lectures and courses focused on the necessity of reporting by using real-world cases and lessons from incidents, which could have been prevented if their associated hazards and risks were reported earlier. One of the cases was about a defect during the rehabilitation of a taxiway surface, which was not reported and led to an aircraft experiencing a flat tyre incident during taxiing procedures. A second case referred to an uncovered utility hole at a construction site which again was not reported and resulted in one site worker getting injured. The massive participation of employees (more than 90%), the positive feedback received and the gradual increase of safety reports were considered as the first and important signs of success. Most of the workers mentioned that they had been convinced to change their behaviour because we, the Safety and Training Department and the Contractor, treated them with integrity, sincerity, a positive attitude and respect for their concerns, ideas and suggestions. Most importantly, the workers stated that after the training courses and during the project execution, they considered that their workplace should be as safe as their homes and decided to behave safely even when nobody was observing their activities.

To promote construction work safety, the Safety Department developed and circulated a "Safety Leaflet" focusing on Dos and Don'ts in such a way that the reader could easily understand the difference between the right and wrong actions and the expectations of the organisation. My intention behind this initiative was not to remind them of their regulatory obligations. By using graphics, I wanted to illustrate what could be the consequences of their activities if performed in ways that would endanger human lives. I provided the main idea and the content, and an experienced graphics company were assigned to develop the leaflet. We paid extra attention to the size and overall quality so that every worker could carry the leaflet during the project activities. Within a year, more than 10,000 pocket-sized copies were distributed at aerodromes where construction works were in progress. We also requested aerodrome managers to make special reference to the content of this leaflet during the monthly meetings of the Safety Committees. Besides, we delivered lectures on safety-related matters (e.g. stress and fatigue management, safety culture, situational awareness, wildlife impact to aviation).

The phase of deployment of the project deliverables into operations was crucial and well-prepared by each aerodrome operator in close coordination with the company's Safety and Compliance Departments. Each aerodrome community knew beforehand about the due date and nature of the upcoming operational changes so that they could plan their actions accordingly and timely. In cases where works were related to parts of manoeuvring areas, special attention was given to the timely communication with all parties involved and the authorities. At least ten days before the scheduled delivery date, a meeting was held locally with the participation of all aerodrome users, including representatives from airline operators, to evaluate the current status, verify regulatory compliance and confirm that all relevant measures were taken for a smooth transition into full operations without an impact on safety.

About a year later, the civil engineering works across all aerodromes were running smoothly and everyone remained focused on the timely delivery of projects without compromising operational safety. I believe that this achievement cannot be attributed to any of the initiatives above separately but is the result of their combination and our collective efforts as well. Thus, my recipe for success includes comprehensive consultation before and during all phases of any project affecting operational safety, regardless of its size and complexity, open communication amongst all parties involved, continuous coordination between aerodrome management and the Contractor's local representatives, customised training for everyone involved with construction works, promotion of reporting culture, continuous monitoring and adherence to regulations.

New System Launched! Did You Consult the Intended End-Users Though?

According to the International Civil Aviation Organisation (ICAO) and the European Aviation Safety Agency (EASA), certified aerodromes must operate an SMS to achieve an acceptable level of safety.[1,2] ICAO and EASA require aerodrome operators to establish mandatory and voluntary reporting systems to facilitate the collection and management of information on actual or potential safety deficiencies, aiming to minimise the loss of human life, property damage, environmental degradation and negative effects on society in general. Voluntary Occurrence Reporting (VOR)[3] refers to the ideal situation where any individual, regardless of his/her relationship with aviation safety, reports anything he/she assumes that endangers personal and/or operational safety. Each aerodrome must have in place a process to facilitate this option. In parallel, there is a regulatory framework for the Mandatory Occurrence Reporting (MOR)[iii] of deficiencies related to the domains of flight operations, aircraft technical, maintenance and repair activities, air navigation services and facilities, and aerodromes and ground services provision.

Actually, a report submitted through the VOR channels can be considered as MOR if the company and authorities deem that it falls under the categories of the latter and poses increased safety risk on operations, although there are cases where an event is recorded parallelly in both the VOR and MOR channels (e.g. accident, specific categories of emergencies, close mishaps). Consequently, mandatory and voluntary reporting policies are an extremely important part of SMS and reflect the behaviour and reaction to identified deficiencies that have or could potentially have an impact on safety. This, in turn, depends on the maturity of reporting culture within each organisation and of every person grounded in the confidence and merit of the reporting process.

When I took over as a safety manager, I noticed that there were various practices regarding reporting, especially amongst those who were actively involved in activities with a direct or indirect impact on safety. Aerodrome staff, although they were aware of their responsibilities, were not convinced about the importance of reporting and its value as a proactive mechanism. They believed they could completely and effectively handle the observed operational deficiencies or human violations. Informal interactions with various employees made me realise that the approach above had been shaped by various factors.

First, there were unfortunate experiences with previous employers who had handled errors or near misses in a punitive manner, without promoting the just culture and the value of reporting. This approach, inevitably, had discouraged the open and honest sharing of safety flaws because the reporter was seen as a troublemaker and subject to disciplinary actions or she/he had committed an error or taken a decision unfavourable to management. The locality also served as a barrier to reporting since, in conjunction with the factor above, the reporter would be tagged as a whistle-blower and harmful for her/his colleagues. A third parameter was the lack of adequate comprehension of the possible consequences of not reporting an identified flaw or a safety occurrence and unawareness that small problems could become larger over time or contribute to major safety issues when combined with other minor deficiencies. Furthermore, I detected a mistaken belief that only managers or supervisors should report accompanied by a lack of encouragement from managers to their staff to report. Additionally, employees were confused about what, how, when, why to report and, perhaps most importantly, there was peer-pressure to keep quiet about any problem at the workplace, safety-related or not, so nobody got into trouble.

Given my strong confidence in the importance of reporting, I arranged training courses for all the aerodrome staff with special attention to the benefits of reporting, process and management of reports and integrity of collected data. The company had already implemented safety reporting software, which had been developed by the Safety and IT Departments. The reporting process and the content of relevant forms were based on the requirements set by ICAO and EASA, including separate forms regarding MOR (a detailed form for accidents/incidents, major hazards and emergency events) and VOR (a simpler form for any other issue). Moreover, MOR forms should be filled and submitted within a specified timeframe. On the contrary, VOR had no limitation on submission time. The initially launched version of the reporting system focused on operational staff who were directly and actively engaged in safety-relevant operational tasks at aerodromes. This initial version did not allow persons who were not working at aerodromes and/or had no direct relation with operational safety to submit a VOR. After one year of implementation, the number of VORs was dramatically low compared with MORs. Actually, only 10% of the total reports received were submitted though the VOR channel, which — at least according to the well-known theory of Heinrich's triangle[4] about the proportions between unreportable, low severity and high severity events — was alarming and a wake-up call for me. I am aware that there is plenty of criticism on the validity of the numbers mentioned in the work of Heinrich, but the underlying concept that unmanaged minor issues can lead over time to major problems is still accepted. We had not set any specific indicator for a fraction of MOR/VOR records, but a 90/10 proportion was a complete inversion of our expectations.

Consequently, we, the Safety Department, were tasked to explore the constraints and potential shortfalls for the insufficiency of VOR records and propose improvements. We organised a survey based on questionnaires administered to the whole workforce and several interviews to capture richer information. The combined findings of our study revealed plenty of interesting practices and perspectives concerning VOR. First, reporters were focused more on MOR because it was directly linked to their responsibilities and represented the bare minimum expected from them regarding safety reporting. Second, staff were focused on reporting regulatory non-compliance issues as required by the national legislation and thought that this was enough to ensure acceptable safety levels. Moreover, voluntary reporting had been given a lower priority by operational staff since it was not a core and central activity and practice; it seemed that the term "voluntary" had been associated somewhat with the perception of uselessness or, in the best case, "good to have". On the more practical side of things, we discovered that the design of the online reporting platform did not facilitate submissions from individuals other than aerodrome staff (e.g. ground handlers, civil aviation authority staff, visitors, passengers) and had created a sense that safety was the responsibility of only a few people. Furthermore, the employees stated that the VOR form was not simple to understand and required considerable time to fill. Thus, staff who intended to submit a VOR and a MOR preferred to invest time in only submitting the MOR form.

Following these unfavourable results from the survey, we reconsidered our initial approach and proceeded with the development of an improved VOR version open to the public and, in parallel with this, encouraged reporting under a blameless and non-punitive environment. The improved web-based system is now user-friendly, simple and accessible to anyone from anywhere. The introductory section of the reporting form highlights the importance of voluntary reporting and corporate commitment regarding the confidentiality of reporting management and the protection of personal data in compliance with the European regulatory framework. After six months of implementation, the VOR number increased remarkably, but the effectiveness and efficiency aspects of the modified reporting are still under evaluation.

The failure with the initial reporting system was an important lesson for me. I learned that human behaviour needs time and continuous efforts to be influenced and changed, especially when you inherit a culturally adverse environment that discourages reporting. To convince people to report even their own mistakes is a goal that can be achieved only if you cultivate a culture of trust amongst all persons involved with activities with a direct or indirect impact on personal safety and consequently on aviation safety. Our initial MOR and VOR systems had met the legislative requirements, the company was compliant and everything should work fine. But the reality proved me wrong. Whatever system you design, the users must see the value and get engaged in its development, operation, maintenance and promotion. Otherwise, without consultation and genuine buy-in, the best of approaches could fail even under the best of intentions and if the design complies with applicable standards.

Notes

1. ICAO (2018). *Safety Management Manual*, 4th ed. International Civil Aviation Organisation, Canada.
2. European Union (2014). Commision Regulation No 139/2014: laying down requirements and administrative procedures related to aerodromes pursuant to Regulation (EC) No 216/2008 of the European Parliament and of the Council. *Official Journal of the European Union*, L44, p.1–34.
3. European Union (2014). Regulation No 376/2014 on the reporting, analysis and follow-up of occurrences in civil aviation. *Official Journal of the European Union*, L122, p.18–43.
4. Heinrich, H.W. (1931). *Industrial Accident Prevention: A Scientific Approach*. New York: McGraw-Hill.

Chapter 17

Just When You Thought You'd Done Enough

Steve Denniss

Contents

These stories refer to the early part of my career as a safety assurance engineer. During this period, I was absorbing new assurance techniques into my personal toolkit, experiencing a range of technologies and industries, and developing my understanding of how and when to apply different approaches to safety assurance.

That Sinking Feeling

Recently qualified, with a BSc in Electrical Engineering, and joining an already established team of analysts, I was very keen to demonstrate my capabilities. I joined the team at an important time. My colleagues had just successfully delivered a safety analysis of an advanced weapon system for the Ministry of Defence (MoD), and we'd been given a lucrative follow-on contract. I was recruited to the team because the system we were going to analyse was larger and more complex than the previous one and we wanted to increase our resources on the job to ensure we wouldn't disappoint the client. The project involved carrying out a safety analysis of a new advanced state-of-the-art electronic underwater guided missile system for the MoD, Royal Navy. Essentially, a torpedo that had its own motor, internal navigation and homing system, and a warhead.

Unlike other industries such as oil and gas, aviation, and rail, weapon systems are designed to be destructive and ultimately have the potential to cause injuries and fatalities rather than prevent them. A definition of safety for a weapon system is related to giving protection to its users. This protection is in two parts. First, to make sure the system works when needed as this will protect the user from enemy forces attack, and, second, to ensure that the people using the system are protected from unsafe conditions of the weapon when in operation. This leaves the question of what the balance between operability and safety should be. When we got our contract and the technical specification for our work, it was clear that the client had considered protection from harm to the users from a malfunction of the system as primary importance. This set the tone for our analysis. In our initial project briefing, it was clear that we needed to consider failure modes, which might result in harm to the users and leave the operability to the experienced weapon system designers. The client had already held a brainstorming session and identified several undesirable events they wanted to avoid if at all possible and had set very stringent targets for these.

There had been some high-profile military conflicts in recent years, and tensions were still high in certain regions. This meant that the defence budget was at a high point, and we were told to do the best job we could. We, as the safety analysts, were given a substantial budget as part of the large sum assigned to this procurement overall. We were to work with the supplier, integrated with the designers, and provide input to ensure that the design they produced met the stringent safety targets. Given the structure of the high-level brief, which was a series of undesirable events, we adopted the Fault Tree methodology to perform the safety analysis.

This was to be an interactive assignment. We needed to work alongside the designers and give them feedback on the design so that together we could ensure the system met the very strict safety targets. This approach meant that we needed to construct a tree and quantify it based on emerging designs, feed the results back to the designers, and update the tree with the updated design. There would be several iterative cycles, and we aimed to improve the safety performance at each cycle. We achieved this way of working by coding the trees in a specially designed fault tree tool we had developed for the previous job. Having coded in the structure of AND and

OR gates, we could then update this as the design evolved. At each stage, we could populate the data with failure rates for each causal event at the lowest level of the tree and provide a probability of the undesirable outcome for that version of the design. I was very keen, along with my colleagues, that we would ensure that the design met and, if possible, exceeded the safety targets and that we demonstrated our worth to the project by improving the safety levels at each design stage.

Having decided and agreed on the approach, the next step was to understand the system and consider the undesirable events in the context of their part in the overall mission. The functional sequence for this weapon system was that once an enemy target had been identified and its characteristics captured and stored in the onboard computer, the torpedo launch sequence would be initiated by an operator. The torpedo tube hatch would open, the torpedo would "pop" out due to the air in the tube and the water rushing in, the torpedo would stabilise in the water due to its aerodynamic design, and approximately six seconds after leaving the tube (enough time for the torpedo to be safely clear of the launching craft), the torpedo engine would start. From this point on, the torpedo would follow its internal navigation system and hopefully strike the target.

Clearly, there are many potential safety issues, for example, the guidance malfunctioning and the torpedo circling round and striking the launch ship, or the torpedo somehow detonating before it had cleared the vicinity of the launch ship. In the initial brainstorming session, the design team had generated a list of credible undesirable outcomes, which we had then captured in fault tree terms as undesirable top events. There were six top events to be analysed. I was assigned the top event "Premature Engine Start". The idea was that if the engine started before the torpedo left the tube, or before it was safely clear of the launch ship (hence the six-second delay), this could cause damage to the ship. The precise engine start sequence involved the collection of several inputs from various processes. These included many circuits monitoring the launch sequence, multiple timers, and status monitors. A final processor board compared the inputs and provided a go or no-go for engine start at a precise time and under acceptable conditions.

I carried out a thorough analysis of the many individual circuits which made up the engine start sequence. It took around six months for each stage of design, of which there were three key design stages. The process involved talking to engineers, studying diagrams and reading system descriptions. I was extremely keen to ensure that my part of the overall weapon system met and exceeded the assigned targets. We were all very much driven by the desire to make sure the events we were analysing would never happen in operation. We successfully achieved a high degree of integration with the designers such that we were able to capture a very detailed model of the safety features of the design and their relationship with the underlying operation. A fact which was to prove extremely valuable later in the project.

Whilst I was focussing on my fault tree, my fellow analysts were similarly focussed on theirs. At key points, we would meet up to share our experiences. It was clear that the designers were pulling out all the stops to design a safe system. There were many check circuits, 2oo3 tests, timing checks, etc. included in the design for each part of the weapon operational process. We had successfully instilled a safety culture across the whole project.

Eventually, we completed the fault trees for the final designs. The probabilities of each of the top events were orders of magnitude better than the requirements. On analysis, I found that for the engine start part of the sequence, there were no cut sets up to order ten that would lead to a premature engine start. A cut set is a combination of events (failure events) which if all true would cause the top event "premature engine start" to occur. Essentially, the probability was extremely low as more than ten failures would have to happen simultaneously for the engine to start prematurely. From a safety perspective, this was a successful outcome, and I was very happy. My tree seemed to demonstrate that the most comprehensive set of fail-safe design features were incorporated in the engine start part of the system. Being the new member of the team, this gave me a great boost with my colleagues as I had seemingly mastered the art of using the fault tree to design out almost every conceivable thing that could go wrong and eliminate the possibility of premature engine start almost entirely. My fault tree was held up as the benchmark for others to achieve in the future. The team moved on to the next assignment and I was given a leading position in the assurance team for the project. Having set our sights on the next job, we took no further interest in the torpedo system, the construction and testing phases of which had begun.

About a year later, the first devices were fitted in the host sub and a sequence of test missions was planned to be carried out in a segregated area of a harbour on the south coast of England. Everyone was very confident in the system design and a successful sequence of tests was anticipated. Safety had been successfully assured by our team so that there were no fears of accidents during the testing. On the first day of the tests, ten attempts were made to test-fire the torpedoes. This involved starting the sequence, homing the weapon-in on a dummy target, arming the dummy warhead, and

firing the torpedo towards the target. On every single test firing, the torpedo popped out of the tube and dropped slowly to the harbour floor. The
engines never started!

We got a call from the client and were requested to attend a meeting. The
failure of the engines to start was obviously not acceptable. I got together
with the designers and we went through the fault tree looking at the safety
circuits. It became clear that in our efforts to demonstrate a highly safe
system, we had designed in a situation where it was almost impossible for
the engine to start at all let alone prematurely. In a classic customer-facing
process plant, or transportation scenario, for example, where harm to the
users was to be avoided, this would have been a costly but highly safe
design. Even in that scenario, it would be beneficial to meet a reasonable
safety target for an acceptable cost, rather than over-engineer the safety,
possibly to the detriment of other features where the budget could have
been spent more effectively. In a military environment where the purpose
is to launch a destructive payload, which inherently carries some risk, making it very unlikely even to achieve its mission was not smart. Gradually, we
re-designed the system to be "less safe" and more "operable" by taking out
some of the safety check circuits. The probability of the top event was now
increased but still kept at an acceptable level and the likelihood of success
was at least reasonable.

Several months later, with a modified design, tests were carried out again.
This time, they were able to successfully start the engines and complete the
series of tests. About a year later, the torpedo system went into service, and
there were no reported incidents of "Premature Engine Start", or, and what
was equally as satisfying, no cases of the engine failing to start. The analysis had been successful in the sense that we could use it to ensure that the
design was safe enough, and perform its function at the same time. I learned
a precious lesson through my involvement in that process. It is ok to aim for
the very highest level of safety; however, safety and operability need to be
considered together to get a balanced outcome. This was a lesson I carried
forward to later projects and indeed has been a key focus of my work in
System Assurance in my subsequent project work.

"Oh Rats:" I Never Thought of That

With a degree in Electrical Engineering behind me, I had completed my
"apprenticeship" so to speak by spending ten years working in the defence

sector. This had allowed me to learn and practice all the assurance techniques such as Fault Tree Analysis, Failure Modes, and Effects Analysis, as well as Spares Provisioning, Logistic Support Analysis, Human Factors, and Life Cycle Costing. These had all been performed to best-practice following the strict defence and military standards and guidelines. So, I was ready to face a new challenge and step out into a more commercial arena to test myself in a slightly less regulated environment. I applied for and got the job of system safety manager for a consultancy responsible for the upgrade of a light railway system in a fast developing and growing part of London.

For this assignment, my overall responsibility was the management and coordination of the complete safety certification process for the project. This role included interfacing with the internal departments responsible for the high-level requirements specification, reviewing and approving supplier's submissions, and liaising with the government department responsible for approval of the railway, which at that time was Her Majesty's Railway Inspectorate (HMRI). The HMRI consisted mainly of several "dyed in the wool" railwaymen from a previous era, and a number of ex-senior ranking officials from the military who had been retired and assigned part-time roles in government departments to keep them busy and "off the streets". This meant that that the inspectorate held a position of power and authority and provided a robust gatekeeper function for new and upgraded railway projects. Being able to convince the man from the Inspectorate that all was well was a key challenge.

A retired Major from the Army had been appointed to oversee our project, and it was my job to convince him that the upgraded railway was safe to operate in passenger service. This was before the current processes of safety management systems and safety cases had been introduced so there was not much in the way of good practice to fall back on. Most railway projects employed someone from the aviation, oil and gas, or other industries who had experience of producing safety arguments. My background in producing assurance submissions in my previous roles in defence projects came in extremely handy. The Major and I built up a good rapport when some of my colleagues found him quite austere and bureaucratic.

At a detailed level, my day-to-day role included the classic system safety management activities. These revolved around management of hazard analyses for each of the technical and operational areas, safety reviews of the designs, and quantitative risk assessments on all system and operational aspects. The project itself included a significant system integration role and interfaces were seen as a major potential risk area. Hazard and safety analysis of the interfaces was included as part of my responsibility. The railway we were to upgrade had been in operation for several years and the success of the service it offered and the area and routes it covered meant that to keep up with the passenger demand, a significant overhaul both of technology and of the procedures and practices were required. The project included the incorporation of a completely new signalling system operated from a purpose-built control room, with modern control systems, displays, and security features. The rolling stock itself was to be replaced with a new fleet of state-of-the-art driverless vehicles, and the operational procedures were to be completely rewritten to support a fast, reliable service to the public. In addition to the technical upgrade, there was a new extension to be built, and several platforms were to be lengthened to support longer trains. I found the experience extremely rewarding as it gave me the opportunity to be involved in an extensive range of technical disciplines and to apply the basic safety tools and techniques on a complex project thus developing my skills and expertise to a new level.

Due to the complexity, the project was divided into several major contracts. Each contract covered one of the new systems, with an integration function within the central project team coordinating and managing interfaces. The project was also divided into several phases, and I was responsible for overseeing the safety reports, justifications, and ultimately safety cases as they evolved and emerged for the individual elements, as well as for the integration function. I performed this role for preliminary design, detailed design, installation, test, and the final safety approval to enter passenger service. As the project progressed through each of the phases, there were design changes. Each of these needed to be assessed for safety impact and the outcome incorporated in the developing safety argument.

Everyone on the project was extremely busy, and we worked long days, nights, and weekends to try to deliver a successful upgrade on time. I became absorbed in the details of the safety cases and this gradually resulted in a myopic view of the hazard mitigation process. There was a drive to demonstrate that risks had been mitigated and that there were no loose ends which may cause concern around the safety process. Everything was focussed on closing-down and completing the safety cases with the minimum of disruption to the project schedule. Despite having several years behind me in safety analysis, I feel that I was still inexperienced in recognising when project deadlines and schedule start to become the key driver and supersede the safety engineer's primary role which is to keep looking for potential hazards and asking the difficult questions. The main focus was on the process not necessarily on capturing all possible risks.

Having divided the scope into its individual disciplines, we set up a safety process for each technical area. The first step in each process was hazard identification. Having identified the hazards, we agreed on mitigations which were assigned to responsible parties to implement. We used a range of hazard identification techniques. These included the standard structured Hazard and Operability (HAZOP) technique of keywords, as well as a list of typical railway hazards. The objective of the process was that the individual discipline hazard databases were complete and independently assessed, and all risks had mitigations assigned. A separate interface hazard analysis was completed along similar lines.

At each stage of the design process, from preliminary design onwards, the hazard databases were checked for consistency and completeness, and the safety case was produced based on the closure of all hazards. At the early stages, closure could consist of a plan to carry out testing or perhaps some routine maintenance check to ensure correct operation in service. The process was very good at ensuring we had followed our plan, applied a generic hazard identification process, kept everything in order, and checked that every item in the log was complete. It was less good at taking a holistic view of whether we had exhaustively captured and designed for hazards associated with the real job in hand.

In my regular meetings with the representative from HMRI, I focussed on presenting a well-drilled team operating a structured process. We had gone through a thorough list of steps and had kept excellent records of everything we'd done. The Major was impressed by the thoroughness of our approach, and I was commended within the project because I was keeping the authorities happy. Whilst the engineers were working on trying to get designs completed, I was taking the pressure off them by demonstrating that the designs were achieving their safety goals and targets. This did give me some leverage within the project, and I was able to call on the support of many of the senior engineers where required to provide technical safety arguments and justifications. On the whole, I managed to convince them that without the safety report and evidence of a safety process, none of their work would matter anyway as we wouldn't be able to put the railway into service. The ongoing good relationship with the HMRI was a useful tool in convincing them to support my efforts. However, the focus was almost always on closing-down what we already had, not on identifying any new hazards which may be emerging.

The design progressed into construction and testing. There were several issues along the way, however, none of these was safety-related, so I managed to keep largely on track with my schedule of safety submissions. Several software issues started to emerge related to the signalling system. These generally resulted in a fail-safe situation, and so they did not impact the safety performance of the railway. It was generally considered that the built-in redundancy and cross-checks within the system would act as mitigating factors in ensuring that safety-critical failures were very unlikely. During the testing phase, we recorded the failures and mapped out a reliability growth of the system, which demonstrated that on the whole, the bug fixing process was successful. During this period, there were no wrong side failures, and the safety analysis began to progress with actual field data supporting the predictions.

Then we started to experience failures of the signalling system, which were of more concern. Some of these, although not actually wrong side, were bordering on serious because of the number of them, the unexpected nature of the faults, the fact that they were widespread throughout large geographical areas, and that they were often intermittent. The number of faults started to mount, and the project delivery was in danger of being severely affected. Then during actual dynamic train testing at low speed, there was the equivalent of a signal passed at danger. Effectively, a train continued moving when it should have stopped at the end of its movement authority. There was a functional investigation and it was found that this could have led to an accident in full operation. It was not an unsafe condition in itself, because other systems would have detected the fault and protected other trains, but it was something which compromised the safety of the railway and should have been identified and mitigated. This brought a stop to testing and the commencement of a physical investigation of the communications systems and bearers.

Investigations of the areas of the faults started to reveal damage to the cables. These were cables which had only recently been installed. It didn't seem that the damage could be put down to vandalism, or construction/maintenance. Quite quickly, someone came up with the cause. It was rats attacking the outer covering of the cables to such an extent that the signalling carrying interior of the cables was affected and compromised, causing commands to be corrupted and erroneous movement of trains to be possible. This answered the question as to why a lot of the faults were inter-mittent. What we had failed to do in our analysis was to consider all aspects of the environment. The light railway was in an area which, had we thought about it properly, was likely to have a high population of rats. Having identi-fied that this was a potential hazard-cause, we re-laid the signalling cables with a combination of troughs and armoured cable to protect the cables and mitigate the issue. Despite having to carry out this relaying exercise, the railway opened on time, and it has continued to operate, to my knowledge at least, without cable faults since opening.

This episode taught me two very important lessons. First of all, I learned that considering the environmental issues of a system is a key factor and one which is often overlooked, whether the environment is extreme heat, dust, sand, or even proximity to people or animals. In later projects, in the Middle East for example, I learned to take account of such environmental issues as drifting sand, extreme heat, and even damage by camels.

The second key lesson and probably the more generic and important one was that there is a risk in focussing on closing-down the hazards in the hazard log as a way of demonstrating achievement of safety. It is important to be mindful at all stages of a project that there may be new hazards which had not previously been identified, and which ultimately may be show stoppers, even though the hazard log may be complete and everything in it closed.

Index

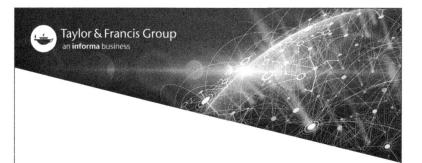

Printed in the United States
By Bookmasters